# Introduction: What is examined on each paper?

Figure 1 outlines how the OCR GCSE Geography A (Geographical Themes) c
is structured. The course consists of three papers and on each paper you m
answer all questions. Make sure you know the content of each paper.

## Paper 1: Living in the UK Today

| What topics are on this paper? | Marks | Assessment overview | |
|---|---|---|---|
| Landscapes of the UK | ≈20 | **1 hour**<br>**60 marks** | **30% of GCSE** |
| People of the UK | ≈20 | | |
| UK Environmental Challenges | ≈20 | | |

**Top tip:** '≈' means roughly 20 marks, rather than exactly. The mark distribution is not split evenly within each paper, so don't get caught out by this on your timings!

## Paper 2: The World Around Us

| What topics are on this paper? | Marks | Assessment overview | |
|---|---|---|---|
| Ecosystems of the Planet | ≈20 | **1 hour**<br>**60 marks** | **30% of GCSE** |
| People of the Planet | ≈20 | | |
| Environmental threats to our Planet | ≈20 | | |

## Paper 3: Geographical Skills

| Section | What topics are on this paper? | Marks | Assessment overview | |
|---|---|---|---|---|
| A | Geographical Skills | ≈40 | **1 hour 30 minutes**<br>**80 marks** | **40% of GCSE** |
| B | Fieldwork Assessment | ≈40 | | |

**Figure 1.1 OCR GCSE Geography A (Geographical Themes) course structure**

N.B. There will be 3 marks for SPaG (spelling, punctuation and grammar) included in the marks for each paper. This will be on either an 8- or 12-mark question and will be indicated next to the question number with a '*'.

There is one mark a minute for Paper 1 and 2 (60 minutes for 60 marks). For Paper 3 however, you have slightly longer (90 minutes for 80 marks), which gives you extra time for some of the skills questions and longer questions.

# 1 How OCR Geography A (Geographical Themes) is assessed in Papers 1, 2 and 3

## Understanding exam questions

To answer an exam question correctly use HUG!

1   **H = HIGHLIGHT the command word**

However, you also need to **understand** what the command word means too, because command words tell you how to answer the questions.

2   **U = Underline the important instructions.**

Once you know the meaning of the command word, underline all the instructions. Sometimes the question is a bit more complex and there are a few instructions. It is important you underline **all** the instructions so that you do **all** that the question is asking for.

Knowing what the question is asking for also involves looking at the tariff (the number of marks the question is worth).

3   **G = Glance back at the question to make sure you are answering it!**

Keep glancing back to the question as you are answering it, to make sure you are continuing to answer the question.

**Tackling complex questions:**

a   The question might require you to study a **figure**, which will usually take the form of a graph, map, photo, table or piece of text. Study it carefully because the question wants you to refer to the **figure** in your answer.

b   The question might require you to refer to a **case study**. This will be one of the ten case studies you have studied (you will find these on pages 29, 32, 37, 40, 43, 53, 56, 60, 64 and 72). Make sure you include case study knowledge such as place names, facts and statistics.

c   The question might require you to write about more than one thing, e.g. environmental **and** political reasons. You must write about both in your answer so give yourself enough time to write about both!

## HUG example

Here is an example of how to HUG a question:

Evaluate whether one piece of primary data collection was successful.   **6 marks**

| Make a judgement considering different factors and using available knowledge/evidence. | All the marks are for **AO3** – Evaluating the success of **one** primary data collection method. Aim for one paragraph on how it was successful, and one on how it was not. |

# OCR
# GCSE (9–1)

# WORKBOOK

# Geography A

Practise your exam skills • Answer questions confidently • Improve your grade

Matthew Fox

HODDER
EDUCATION
LEARN MORE

# Contents

| Command word | Definition | Example | Tariff |
|---|---|---|---|
| Calculate | Work out a numerical answer | Calculate the mean life expectancy. [1] | Low tariff (1–3 marks) |
| Compare | Identify similarities and differences | Compare how continentality influences the weather in Nottingham and Plymouth. [2] | |
| Define | State the meaning | Define a drought. [1] | |
| Identify | Recognise, or name a feature | Identify **three** pieces of evidence that suggest this is a tundra ecosystem by annotating onto the photo. [3] | |
| Outline | A description setting out main points | Outline **three** features of plants in tropical grasslands. [3] | |
| State | Express in precise terms | State **two** warm water coral reefs that can be found near Central America. [2] | |
| Describe | Give an account, including all the relevant characteristics/events | Describe the differences between the top 10 UK imports and exports in 2017. [3] | |
| Discuss | Give an account that addresses a range of ideas and arguments | Discuss which stage an LIDC or an EDC you have studied has reached on Rostow's model. [6] | Medium tariff (4–8 marks), although these are also used in some 1–3 mark questions |
| Explain | Provide the purposes or reasons | Explain the economic and environmental consequences of development in your chosen place/region. [8] | |
| Suggest | Give possible alternatives, produce/put forward an idea | Suggest how aid can both promote and hinder development. [4] | |
| Analyse | Separate information into components and identify their characteristics. Discuss the pros and cons of a topic or argument and make a reasoned comment | Suggest a conclusion that the students might reach for the enquiry question 'How do patterns of solar panel usage vary?' Analyse the evidence from the table to explain how you have reached that conclusion. [6] | High tariff (6–12 marks) |
| Assess | Offer a reasoned judgement informed by relevant facts | Assess the management of the flood event at **two** different scales. [6] | |
| Evaluate | Make a judgement considering different factors and using available knowledge/evidence | Evaluate the sustainability of strategies used in your chosen city to overcome **one** challenge. [12] | |
| Examine | Investigate/scrutinise carefully, or in detail | Examine the impact of geomorphic processes on the formation of landforms in your chosen river basin. [8] | |
| How far/ to what extent | Make a judgement by considering the arguments for and against. Justify your decision | 'Humans have had a more significant influence on your chosen coastal landscape than geomorphic processes.' To what extent do you agree? [12] | |

**Figure 1.2 The command words OCR A has used since the course was introduced, what they mean, and what types of questions they have been used on**

## Assessment Objectives (AO)

The **Assessment Objectives** are what the examiner is looking for when marking your answer. There are four of them (see Figure 1.3).

| Assessment Objective | Description | What command word might be used? |
|---|---|---|
| 1 | Demonstrate **knowledge**, e.g. locations, places, processes, environments and different scales. | Define, Describe, Explain, Identify, Outline, State |
| 2 | Demonstrate **geographical understanding** of concepts and the inter-relationship between places, environments and processes. | Discuss, Explain, Suggest |
| 3 | Apply knowledge and understanding to **interpret, analyse and evaluate** information, issues and to make judgements. | Analyse, Assess, Discuss, Evaluate, Examine, To what extent |
| 4 | Use a variety of **skills and techniques** to investigate questions and issues, e.g. graphs and maps. | Calculate, Compare, Describe |

**Figure 1.3 Assessment Objectives**

Longer tariff questions (6-, 8- and 12-mark questions) have been colour coded in this book according to their AO colour on Figure 1.3.

Low-tariff questions (1–4 marks) will only examine one AO. Longer questions may have multiple AOs (4–12 marks).

## Spelling, punctuation and grammar marks

Each paper has 3 marks for spelling, punctuation and grammar (SPaG). The examiner is looking for the following three things:

1   Spelling and punctuation – be accurate!

2   Grammar and meaning – ensure your answer makes sense and is logically structured!

3   Key geographical vocabulary – include a wide range of relevant key vocabulary!

# Tackling skills questions

Twenty-five per cent of marks in your Geography GCSE will assess AO4 (skills) – 7 per cent of these marks will be in Paper 1 and 2, 18 per cent in Paper 3. Skills questions involve using graphs, reading maps, interpreting photos, pictures and text, and completing calculations.

## Tackling graphs

| Graphs/charts | What it is used to show | Example | Can you identify? |
|---|---|---|---|
| Bar graph (horizontal, vertical and divided) | Uses bars to represent data in categories | Largest imports and exports (page 34), amount of electricity generated from renewable energy sources in the UK (page 46) | |
| Histogram (equal class interval) | Uses bars to represent the frequency distribution of numerical data | Migration numbers split into age brackets (page 8) | |
| Line graph | Uses a line to represent changes over time | $CO_2$ levels over history (page 9), Demographic Transition Model (page 38), UK electricity generation by source over time (page 48) | |
| Scatter graph | Uses points plotted on a graph to show the relationship between two variables | Relationship between GNI/capita and life expectancy (page 59) | |
| Dispersion graph | Uses points to show the range of data and the distribution of each piece of data within that range | Pebble size (page 85) | |
| Pie chart | Uses a circle to show the percentage of each category of data (adds up to 100%) | Percentage share of gross disposable income in UK (page 35) | |
| Climate graph | Uses a line to show temperature and bars to show precipitation | Climate for a location (page 52) | |
| Proportional symbols | Uses symbols proportional to the value they represent | Proportional circles for a country's population size | |
| Pictograms | Uses pictures to represent values | Traffic count for different types of vehicles | |
| Cross-sections | Uses a line to show the shape you get when you cut straight through an object | The profile of a V-shaped valley | |
| Population pyramids | Uses horizontal bars to show the age–sex distribution of a population | Population pyramid for India (page 83) | |
| Radial graphs | Uses a line to show multivariate data on axes starting from the same point | Environmental Quality Survey results for different categories | |
| Rose charts | Uses equal segments for each category on axes starting from the same point | Average wind direction over a given time (each segment represents orientation) | |

**Figure 1.4 Types of graphs you need to be familiar with**

Graphs will be used in your exam to test the following skills:

- **Completing graphs** by adding data to them.
- **Reading and interpreting data** from the graph.
- **Describing the pattern or trend** of the data on the graph.

> Tick when you have revised and are familiar with what each type of graph looks like.

Here is a quick reminder of what every good graph should contain:

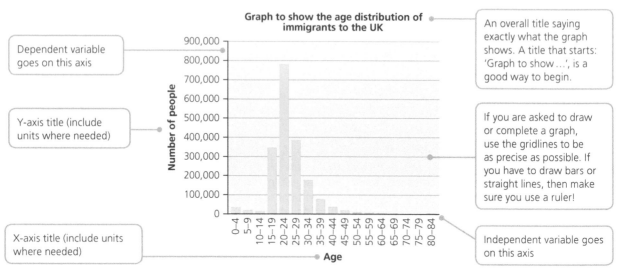

Dependent variable goes on this axis

Y-axis title (include units where needed)

X-axis title (include units where needed)

An overall title saying exactly what the graph shows. A title that starts: 'Graph to show...', is a good way to begin.

If you are asked to draw or complete a graph, use the gridlines to be as precise as possible. If you have to draw bars or straight lines, then make sure you use a ruler!

Independent variable goes on this axis

**Figure 1.5 What a good graph should contain**

## Completing, reading and interpreting graphs

**1** Study **Figure 1.6** which shows the proportion of urban populations living in slums.

a Complete the bar graph to show the data for the proportion of urban populations living in Oceania and Western Asia.

**2 marks**

|  | North Africa | Sub-Saharan Africa | Latin America/ Caribbean | Eastern Asia | Southern Asia | South East Asia | Western Asia | Oceania |
|---|---|---|---|---|---|---|---|---|
| 1990 | 33 | 69 | 34 | 44 | 58 | 52 | 23 | 24 |
| 2010 | 12 | 62 | 24 | 27 | 36 | 31 | 26 | 25 |

**Figure 1.6 A table showing the proportion of urban populations living in slums**

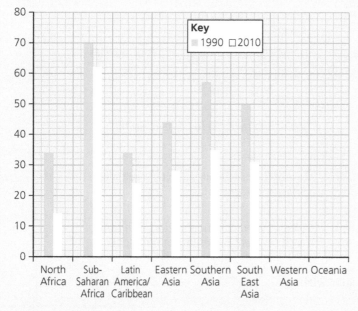

**Key**
■ 1990 □ 2010

**Top tip:** Make it clear which bar is 1990 and 2010 by replicating the shading in the key.

**Figure 1.7 A bar graph showing the proportion of urban populations living in slums**

b Which region has a decrease of 17 per cent of the proportion of their urban population living in urban slums between 1990 and 2010?

A North Africa

C Eastern Asia

B Latin America/Caribbean

D Western Asia

Write the correct letter in the box.

**1 mark**

## Describing patterns and trends on graphs

For Question 2, below, you need to describe the **pattern** (i.e. the overall change over the whole 400,000 years). 'GSD' will help you do this:

General pattern – state the overall pattern, e.g. There is an uneven pattern…

Specific points – go into specifics about the pattern. (Hint: use good descriptive adjectives and see if you can spot an anomaly.)

Data – quote data from the graph to illustrate the pattern you have identified.

**Top tip:** Start 'broad' with your description and then progressively narrow down into details.

**Top tip:** Include good descriptive adjectives.

If the line is increasing/decreasing, use an adjective to comment on the rate: **slowly, steadily, rapidly, exponentially.**

If the line is going up and down: **fluctuating.**

Descriptions of graphs are usually worth 3 or 4 marks; 3-mark questions need three different points about the pattern *if* the question asks for **data**, otherwise the third mark is a communication mark. For 4-mark questions there is *always* a communication mark. Using good descriptive adjectives and writing in a logical order is key to getting the communication mark. Follow 'GSD' (above) for a clear and logical order.

---

2 Study **Figure 1.8** which shows $CO_2$ levels in the atmosphere over the last 400,000 years.

Describe the <u>pattern</u> of $CO_2$ levels in the atmosphere over the last 400,000 years. **4 marks**

**Top tip:** Watch out for if the question asks for **pattern** or **trend**. If you are asked to describe the **trend**, then you need to divide up the graph into chunks and describe each 'chunk' separately. Good descriptive adjectives to describe each trend are crucial.

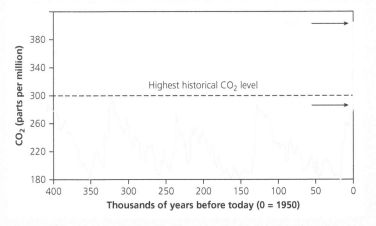

Figure 1.8 $CO_2$ levels in the atmosphere over the last 400,000 years.

**General pattern:** *Overall* … ...........................................................................................

**Specific points:** ...........................................................................................

...........................................................................................

**Data:** ...........................................................................................

...........................................................................................

---

## Want more?

Answer Question 2 again by replacing **pattern** with **trend**. Think carefully about how you would answer it differently. This time it is worth 3 marks and the question wants data (so aim for three trends).

# Tackling maps

| Map | What it is used to show | Example | Can you identify? |
|---|---|---|---|
| Atlas map | Atlases contain a collection of maps including political maps (with national boundaries and capitals) and physical maps (with landscape features such as mountains, rivers, towns and transport routes) | Map of UK with key towns | |
| OS map (1:50,000 and 1:25,000) | A detailed map of the British landscape | Landscape of Kirkby Moor (page 75) | |
| Base map | A map with only essential information to be used as a background setting | Outline of countries | |
| Choropleth map | A map with areas shaded or patterned in proportion to the value of the particular quantity in that area | Mean UK earnings by constituency | |
| Isoline map | A map with lines on to link areas that share the same value | Temperatures across the UK | |
| Flow line map | A map with arrows showing exact path of movement of objects between different areas | Direction and volume of travellers on London tube (page 76) | |
| Desire-line map | A map with straight lines of proportional thickness representing strength of desire to move (rather than exact path) of people or goods between different areas | Movement of trade between countries | |
| Sphere of influence map | A map showing the region of which something has control | Regions people migrate/ commute from to an urban area | |
| Thematic map | A map emphasising a theme or topic | Global distribution of coral reefs (page 13) or tropical rainforests (page 52) | |
| Route map | A map that displays roads and transport links | Travel route to a location | |
| Sketch map | A roughly drawn map showing only the main features of an area | Shows key features in a settlement | |

**Figure 1.9 Types of maps you need to be familiar with**

Maps will be used in your exam to test the following skills:
- **reading and interpreting** information from the map
- **describing the pattern or distribution** of the data on the map.

> Tick when you have revised and are familiar with what each type of map looks like.

## Reading and interpreting information from a map

Every map should contain a north arrow, scale and a key. These will help you with a sense of direction, distance and understanding the key information on the map.

If you have an Ordnance Survey (OS) map in the exam, it will either be at 1:50,000 (i.e. 1cm = 500m in real life) or 1:25,000 (i.e. 1cm = 250m in real life) scales. You may need to use four- or six-figure grid references to locate specific features, draw sketch maps, use scales to measure distances, or use contour lines and spot heights to draw cross-sections.

## Four- and six-figure grid references

A four-figure grid reference identifies a square on a map. First use the eastings to **go along the corridor** until you come to the bottom left-hand corner of the square. Write the two-figure number down. Then use the northings to go **up the stairs** until you find the same corner. Write down this two-figure number after your first two. This is your four-figure grid reference, e.g. the blue shaded grid square in **Figure 1.10** is 6233.

A six-figure grid reference identifies a more exact point on a map such as the dark blue mini-square in **Figure 1.10**. Write out your four-figure grid reference as follows: 62_ 33_

The grid square has been divided up into 100 tiny squares with 10 squares along each side (you will have to imagine this on an actual OS map). Go along the corridor and up the stairs again to work out the extra numbers. You need to put them into a six-figure grid reference as follows: 625 333.

For further practice with OS maps go to pages 74–5.

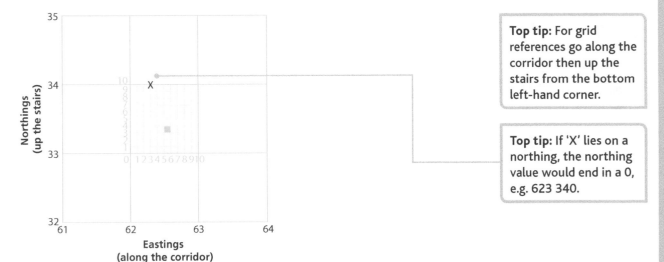

**Top tip:** For grid references go along the corridor then up the stairs from the bottom left-hand corner.

**Top tip:** If 'X' lies on a northing, the northing value would end in a 0, e.g. 623 340.

**Figure 1.10 Model grid for an OS map**

## Measuring distances

To measure a straight-line distance, place a piece of paper between the two points and mark onto it the start and end point. Then place this paper next to the scale to work out the distance.

However, if the distance is curved, mark the first section on your piece of paper, then use a pencil to pivot the paper, so that it follows along the next straight section. Keep doing this until you get to the end. **Figure 1.11** shows how this is done.

## GIS (Geographical Information Systems)

Geographical Information Systems contain geographical data that has a location, which is then mapped. You can then analyse this data to identify patterns. Initially GIS start with a base map, then layers of information are added on top. You may be asked to interpret maps produced by GIS.

Let's practise some questions …

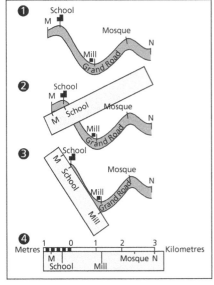

Curved distances

**Figure 1.11 Measuring curved distances**

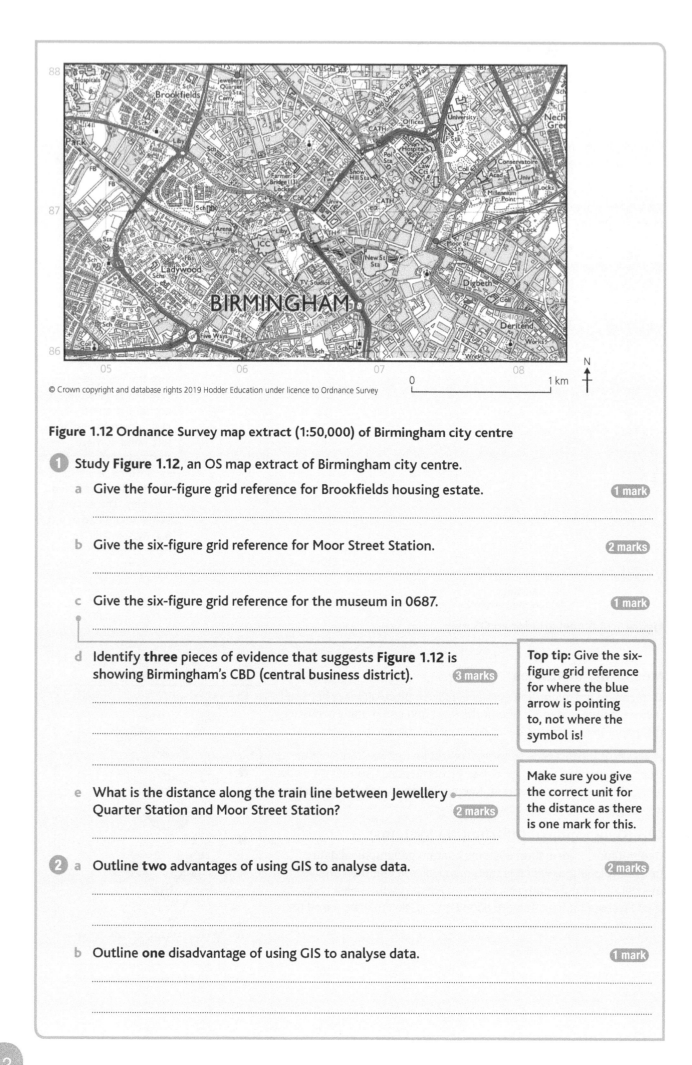

© Crown copyright and database rights 2019 Hodder Education under licence to Ordnance Survey

**Figure 1.12** Ordnance Survey map extract (1:50,000) of Birmingham city centre

**1** Study **Figure 1.12**, an OS map extract of Birmingham city centre.

a Give the four-figure grid reference for Brookfields housing estate.　　1 mark

.............................................................................................................

b Give the six-figure grid reference for Moor Street Station.　　2 marks

.............................................................................................................

c Give the six-figure grid reference for the museum in 0687.　　1 mark

.............................................................................................................

d Identify **three** pieces of evidence that suggests **Figure 1.12** is showing Birmingham's CBD (central business district).　　3 marks

.............................................................................................................

.............................................................................................................

.............................................................................................................

**Top tip:** Give the six-figure grid reference for where the blue arrow is pointing to, not where the symbol is!

e What is the distance along the train line between Jewellery Quarter Station and Moor Street Station?　　2 marks

.............................................................................................................

Make sure you give the correct unit for the distance as there is one mark for this.

**2** a Outline **two** advantages of using GIS to analyse data.　　2 marks

.............................................................................................................

.............................................................................................................

b Outline **one** disadvantage of using GIS to analyse data.　　1 mark

.............................................................................................................

.............................................................................................................

## What makes a good map description?

If you need to describe a pattern or distribution, the following words may help:

**Concentrated**  Information on a map is clustered into small groups.

**Sparse**  Information on a map is spread out.

**Linear**  Information on a map is spread out along lines.

**Random**  Information on a map has no clear pattern and is irregularly spaced.

**Regular**  Information on a map is evenly spaced.

Aim to be as precise as possible with your description:

- Use a compass direction to indicate where something is on the map.
- To be even more precise, provide the distance a point is from another point, either using the scale provided on the map or lines of latitude and longitude.

**Top tip:** You are expected to use whatever information has been added to the map, e.g. lines of latitude, a key or a scale.

**Figure 1.13 The global distribution of coral reefs.**

③ Study **Figure 1.13** which shows the global distribution of coral reefs.

Describe the global distribution of coral reefs in **Figure 1.13**.

**4 marks**

...................................................................................................................................

...................................................................................................................................

...................................................................................................................................

...................................................................................................................................

...................................................................................................................................

**Key terms**

Use the following key vocabulary and techniques to help answer question d below: **concentrated, compass direction, lines of latitude.**

**Top tip:** Use 'GSD' for this description (see page 9). When you get to 'S' for maps you need to contrast where values are high against where they are low.

## Want more?

You can find other types of map to practise describing on pages 12, 42, 51, 52, 75 and 76.

# Tackling photos

There can be a range of ways you may be asked to use the photo in a question. These include:

- Identifying features/characteristics in the photo.
- Describing the photo.
- Annotating and labelling characteristics in the photo.
- Interpreting the photo.

Whatever the question, it is crucial you study the photo carefully.

## Identifying features on a photo

You might be required to identify a feature or characteristic in the photo.

1 Study **Figure 1.14** which shows Whiterocks beach, Portrush, Northern Ireland.

   a  **Identify the landform in the photograph.**

   A   Arch          C   Cave

   B   Spit          D   Waterfall

   Write the correct letter in the box.

**Figure 1.14 Whiterocks beach, Portrush, Northern Ireland**

## Describing characteristics of a photo

Some questions require you to describe characteristics in a photo. This requires spotting evidence in the photo that you can use in your answer. You should use words such as **foreground**, **midground** and **background** to help you describe where the characteristics are in the photo.

**Top tip:** Pick out little details, e.g. the undercut at the base of the arch.

   b  **Describe the features of the landform in Figure 1.15.** `3 marks`

   .................................................................................................................................................

   .................................................................................................................................................

   .................................................................................................................................................

   .................................................................................................................................................

## Annotating and labelling photos

Sometimes questions require you to annotate or label features on a photo.

Read the examples of a label (red box) for Question 2a and annotation (blue box) for Question 2b on page 15.

**Key terms**

An **annotation** provides an explanation of the feature in a photo (a sentence).

A **label** identifies a feature in a photo (one or two words).

**2** Study **Figure 1.15** which shows a slum in an LIDC.

a Identify **two** pieces of evidence in the photo that this is a slum, using labels. **2 marks**

b Outline **two** social consequences of rapid urbanisation in the photo of this slum, using annotations. **2 marks**

One label (red) and one annotation (blue) have been done for you. Add two more of each.

Corrugated iron roofs

Water from the river is used for washing as no infrastructure has been created for providing clean, running water.

**Top tip:** Your arrow must be drawn to the exact place on the photo. Nearby is not close enough!

**Figure 1.15 A slum in an LIDC**

## Interpreting the photo

Some questions require you to use a combination of the photo and your own knowledge, like the example below.

c **Using Figure 1.15** and your own knowledge, suggest **two** environmental consequences of rapid urbanisation. **2 marks**

'**Suggest**' means you need to propose a possible answer based on what you can see, using your knowledge of the environmental consequences of rapid urbanisation.

.......................................................................................................................

.......................................................................................................................

.......................................................................................................................

.......................................................................................................................

## Tackling numerical questions

Some questions require you to do short calculations to process geographical data in a table, or on a graph or map. Although they are only worth a few marks, it's a good way to get these marks as the questions are straightforward! The command word will usually be 'calculate' and you may use a calculator to help you. You need to be able to calculate the **median**, **mean**, **mode**, **range**, **interquartile range**, and **percentage**. Read the examples below in **Figure 1.16**.

| Site | A | B | C | D | E | F | G | H | I |
|---|---|---|---|---|---|---|---|---|---|
| Number of wind turbines | 30 | 2 | 25 | 67 | 160 | 48 | 73 | 48 | 100 |
| Cost (millions) | £123 | £4 | £90 | £1,500 | £2,000 | £300 | £736 | £396 | £900 |

**Figure 1.16 The cost and size of nine wind farms in the UK**

## Averages

**1** Study **Figure 1.16** which shows the cost and size of wind farms in the UK.

Top tip: If the question asks you to show your working, you may pick up a mark for it, even if you get the answer wrong.

a) Calculate the **mean**, **median**, and **mode** of the number of wind turbines in the UK. **Show your working.**

**Mean** = add up all the values in the data set and divide by the number of values. For example:

$$\frac{(30 + 2 + 25 + 67 + 160 + 48 + 73 + 48 + 100)}{9} = \frac{553}{9} = 61.4 \text{ (rounded to 1 d.p.)}$$

**Median** = arrange the data in order (lowest to highest) and the median is the middle value in that rank order. For example:

2, 25, 30, 48, ⟨48⟩ 67, 73, 100, 160 = **48**

**Mode** = the most frequent value in the data set. For example:

2, 25, 30, ⟨48⟩⟨48⟩ 67, 73, 100, 160 = **48**

## Ranges

b) Calculate the **range** and **interquartile range** of the cost of wind turbines in the UK. Show your working.

**Range** = the difference between the highest and lowest values in the data set. For example:

Top tip: Don't forget to add the correct unit!

4, 90, 123, 300, 396, 736, 900, 1,500, 2,000 = 2,000 − 4 = £1,960 (million)

**Interquartile range** = the difference between the three-quarters and one-quarter value. First work out the median, then if the one-quarter and three-quarters values lie between two numbers, add them together and divide by 2. For example:

4, 90, 123, 300, ⟨396⟩ 736, 900, 1,500, 2,000 = 1,200 − 106.5 = £1,093.5 (million)

median

1st quartile = $\frac{(90 + 123)}{2}$ = 106.5     3rd quartile = $\frac{(900 + 1,500)}{2}$ = 1,200

## Percentages

c) Calculate the **percentage** of wind farms located in England. Show your working.

**Percentage** = a proportion of a whole.

There are nine wind farms in total. Six are in England.

i) Divide the number in England (6) by the total number (9): = 6/9

ii) Multiply this number by 100.

= (6/9) × 100 = 66.67%
= 66.7% (rounded to 1 d.p.)

## Going further

Sometimes you are asked to calculate the percentage increase/decrease.

For example: Calculate the percentage increase in megacities between 2016 (31) and 2030 (41).

**1** Work out the **difference** (increase) between the two numbers you are comparing: 41 − 31 = 10.

**2** Then divide the increase by the original number and multiply the answer by 100: (10/31) × 100 = 32.3%.

N.B. If your answer is a negative number, then this is a percentage decrease.

| | 2008 | 2009 | 2010 | 2011 | 2012 | 2013 | 2014 | 2015 | 2016 | 2017 |
|---|---|---|---|---|---|---|---|---|---|---|
| Total no. of tropical storms | 16 | 9 | 19 | 19 | 19 | 14 | 8 | 11 | 15 | 17 |
| Of which became hurricanes | 8 | 3 | 12 | 7 | 10 | 2 | 6 | 4 | 7 | 10 |

**Figure 1.17** Tropical storms recorded in the Atlantic Ocean, 2008–2017

1. Study **Figure 1.17** which shows the number of tropical storms in the Atlantic Ocean, 2008–2017.

   a Calculate the **mean** value of tropical storms from 2008–2017. Show your working. **2 marks**

   b Calculate the **median** value of tropical storms from 2008–2017. Show your working. **2 marks**

   c Calculate the **modal** value of tropical storms from 2008–2017. Show your working. **2 marks**

   d Calculate the **range** of hurricanes from 2008–2017. Show your working. **2 marks**

   e Calculate the **interquartile range** of hurricanes from 2008–2017. Show your working. **2 marks**

   f Calculate the **percentage** of tropical storms that become hurricanes from 2008–2017. Show your working. **3 marks**

# Tackling the longer questions – 6-mark questions

The first type of longer-tariff questions are worth 6-marks. Typically, there are:

■ **One** in Paper 1 (for **one** of the three topics)*.
■ **One** in Paper 2 (for **one** of the three topics)*.
■ **Three** in Paper 3 (**one** in Section A, **two** in Section B).

* Typically, on Paper 1 and 2 there will be one 6-, 8-, and 12-mark question, each on a different topic. Aim to write your answer in six minutes. No need for an introduction – dive straight in!

## How are 6-mark questions examined?

Six-mark questions are marked using levels – the quality of your answer is matched up with the level it fits best with. Six-mark questions have three levels, with the highest being Level 3 (**thorough**). The table below highlights what is required to be 'thorough' for each of the three relevant AOs for a 6-mark question.

> **'Accurate'** – the key word to be 'thorough'!

> **'Range'** – aim for at least three developed points for a 6-mark question.

> **'Supported'** – you need to back up your evaluation/judgements with evidence.

| | AO1 (knowledge) | AO2 (understanding) | AO3 (application) |
|---|---|---|---|
| **Thorough** (5–6 marks) | A **range** of **accurate** knowledge, relevant to the question. | A **range** of **accurate** understanding relevant to the question. | **Accurate** interpretation/analysis through the application of relevant knowledge and understanding. **Supported** evaluation/judgement through the application of relevant knowledge and understanding. |
| **Reasonable** (3–4 marks) | **Some** knowledge relevant to the question. | **Some** understanding relevant to the question. | **Some accuracy** in interpretation/analysis through the application of some relevant knowledge and understanding. **Partially supported** evaluation/judgement through the application of some relevant knowledge and understanding. |
| **Basic** (1–2 marks) | **Limited** knowledge relevant to the question. | **Limited** understanding relevant to the topic or question. | **Limited accuracy** in interpretation/analysis through lack of application of relevant knowledge and understanding. **Unsupported** evaluation/judgement through lack of application of knowledge and understanding. |

To be 'thorough', it is important you carefully read the whole question to determine exactly what the examiner is looking for – **HUG** it!

## Command words

Command words used in 6-mark questions have included:

Discuss    Describe    Explain    Assess    Examine    Evaluate

Six-mark questions in Paper 1 and 2 usually have 2 or 3 marks for AO2, and then the command word determines whether the other 3 or 4 marks are for AO1, AO2, or AO3. Let's consider a couple of examples.

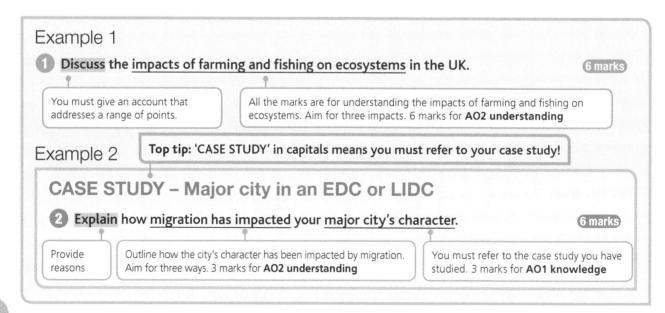

### Example 1

**1  Discuss** the **impacts of farming and fishing on ecosystems** in the UK.    *6 marks*

> You must give an account that addresses a range of points.

> All the marks are for understanding the impacts of farming and fishing on ecosystems. Aim for three impacts. 6 marks for **AO2 understanding**

### Example 2    **Top tip: 'CASE STUDY' in capitals means you must refer to your case study!**

#### CASE STUDY – Major city in an EDC or LIDC

**2  Explain** how **migration has impacted** your **major city's character**.    *6 marks*

> Provide reasons

> Outline how the city's character has been impacted by migration. Aim for three ways. 3 marks for **AO2 understanding**

> You must refer to the case study you have studied. 3 marks for **AO1 knowledge**

## Developing your sentences

To be 'thorough', you need to develop your sentences. Connectives will help you do this as they link simple ideas together. When you make a point, ask yourself: 'what is the significance of this?' This will help you to explain the consequences of your previous statement and turn your sentence into a longer and more developed sentence. Here are some examples:

| Connectives to explain | Developing your explanation even further ... |
|---|---|
| As a consequence of ... | In addition ... |
| ... consequently ... | As well as ... |
| As a result of ... | What is more ... |
| An effect of ... | Also ... |
| ... because ... | Furthermore ... |
| ... so ... | Moreover ... |

**Figure 1.19 Connectives for developing sentences**

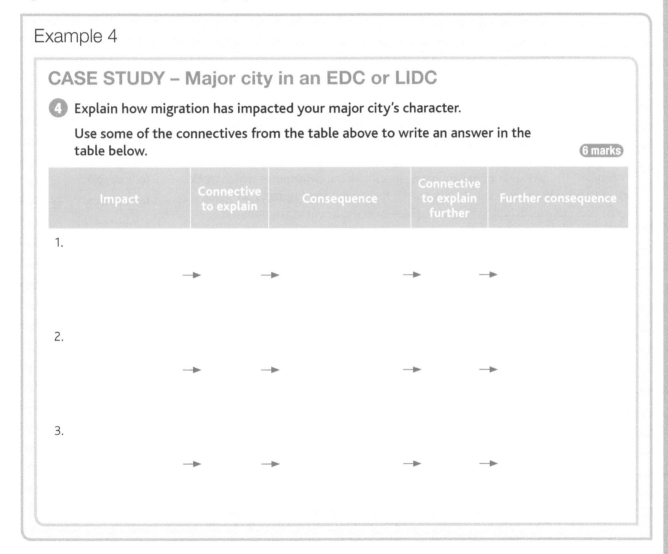

### Example 4

#### CASE STUDY – Major city in an EDC or LIDC

④ Explain how migration has impacted your major city's character.

Use some of the connectives from the table above to write an answer in the table below.

**6 marks**

| | Impact | Connective to explain | Consequence | Connective to explain further | Further consequence |
|---|---|---|---|---|---|
| 1. | | → | → | → | → |
| 2. | | → | → | → | → |
| 3. | | → | → | → | → |

## Dealing with sources

The final type of 6-mark question is where a source is provided. You might be provided with a photo, map, graph or text, which you will need to refer to in your answer. It will have been chosen carefully to include relevant information that would be useful in your answer. You will not get into Level 3, 'thorough', unless you refer to the source!

These questions are particularly common on Paper 3 with the synoptic questions (see pages 61 and 83 for examples of this type of question).

# Tackling the longer questions – 8-mark questions

The second type of longer-tariff questions are 8-mark questions. Typically, there is:

- **One** in Paper 1 (for **one** of the three topics)*.
- **One** in Paper 2 (for **one** of the three topics)*.
- **Two** in Paper 3 (for **one** in Section A, **one** in Section B). In Section B it will have an extra 3 marks for spelling, punctuation and grammar (SPaG).

* Typically, on Paper 1 and 2 there will be one 6-, 8-, and 12-mark question, each on a different topic.

Aim to write your answer in eight or nine minutes. Generally, there is no need for an introductory sentence, unless you need to outline the case study you are referring to. Otherwise dive straight in!

## How are 8-mark questions examined?

Like 6-mark questions, there are the same three levels (see page 18), divided up as follows:

| Level | Level descriptor | Number of marks |
|-------|------------------|-----------------|
| 3 | Thorough | (6–8 marks) |
| 2 | Reasonable | (3–5 marks) |
| 1 | Basic | (1–2 marks) |

**Figure 1.20 Levels of response for 8-mark questions**

There are two extra things the examiner is looking for in 8-mark questions.

- **Place-specific detail**: When told to refer to a case study in your answer, place-specific details must be included to get above 4 marks. For Level 3, 'thorough', the amount of relevant place-specific detail determines which mark you will get within the level. So it is crucial for top marks! **Figure 1.21** gives some connectives to help you include place-specific detail in your answer.
- **Structure**: Level 3, 'thorough', requires a clear and logical structure. Your reasoning needs to be well developed and relevant. The use of paragraphs (probably two and a conclusion if you need to come to a judgement) is therefore important. For more advice on structure, see page 29.

| Evidence connectives |
|---|
| ... as exemplified by ... |
| ... for example ... |
| ... for instance ... |
| This was seen in ... |
| ... such as |

**Figure 1.21 Evidence connectives**

## Command words

Command words used in 8-mark questions have included:

| Explain | To what extent do you agree? | Analyse | Examine | Evaluate |
|---|---|---|---|---|

**Figure 1.22 Command words for 8-mark questions**

You will notice in **Figure 1.22** the command words are becoming increasingly evaluative, compared to lower-tariff questions. Typically, at least 4 of the 8 marks are given to AO3 (analysis and evaluation).

## Evaluative connectives

To pick up AO3 marks, evaluative connectives will help you to make evaluative points. **Figure 1.23** contains a variety of examples of these.

| Evaluative connectives | |
|---|---|
| However ... | ... whereas ... |
| Nevertheless ... | Conversely ... |
| On the one hand ... ... on the other hand ... | As opposed to ... |
| Alternatively ... | Rather than ... |
| In contrast ... | On the contrary ... |

**Figure 1.23 Evaluative connectives**

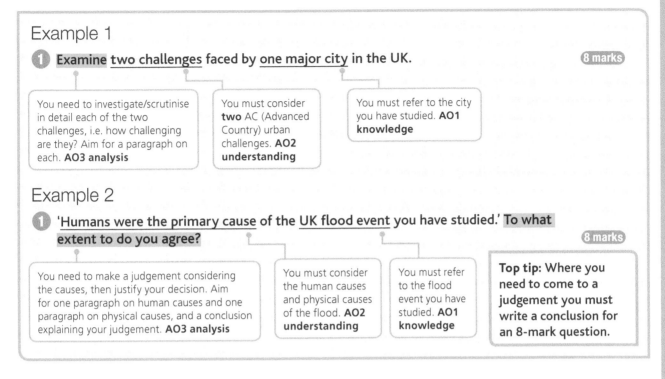

## Example 1

**1** **Examine** <u>two challenges</u> faced by <u>one major city</u> in the UK.

**8 marks**

You need to investigate/scrutinise in detail each of the two challenges, i.e. how challenging are they? Aim for a paragraph on each. **AO3 analysis**

You must consider **two** AC (Advanced Country) urban challenges. **AO2 understanding**

You must refer to the city you have studied. **AO1 knowledge**

## Example 2

**1** '<u>Humans were the primary cause</u> of the <u>UK flood event</u> you have studied.' To what extent to do you agree?

**8 marks**

You need to make a judgement considering the causes, then justify your decision. Aim for one paragraph on human causes and one paragraph on physical causes, and a conclusion explaining your judgement. **AO3 analysis**

You must consider the human causes and physical causes of the flood. **AO2 understanding**

You must refer to the flood event you have studied. **AO1 knowledge**

**Top tip:** Where you need to come to a judgement you must write a conclusion for an 8-mark question.

## Opinion line

To answer a 'to what extent/how far do you agree?' question, you need to come to a judgement, and therefore you need a conclusion. Using the 'opinion line' in **Figure 1.24** will help you come to a decision.

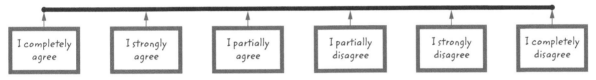

| I completely agree | I strongly agree | I partially agree | I partially disagree | I strongly disagree | I completely disagree |

**Figure 1.25 An 'opinion line' to help you state your decision**

Even if you completely agree/disagree with the argument, you should **always** present both sides of an argument and then make a judgement. To do this, work out where you place yourself on the 'opinion line'. Don't 'sit on the fence' and go for the middle. Choose one side.

Then you need to justify your decision. The following phrases will help you:

I completely agree/disagree ...

I agree/disagree ...

I partially agree/disagree ...

... as the evidence is clear that ...

... as the evidence suggests ...

... as the evidence is inconclusive ...

### Activity

For Example 2 above, using the case study you studied, mark on the 'opinion line' where you would place yourself. Then write a conclusion justifying your decision on the answer lines below. Use the phrases above to help you.

..................................................................................................................................................

..................................................................................................................................................

..................................................................................................................................................

..................................................................................................................................................

# Tackling the longer questions – 12-mark questions

The longest-tariff questions are worth 12 marks. Typically, there is:
- **one** in Paper 1 (**one** of the three topics)*
- **one** in Paper 2 (**one** of the three topics)*

* Typically, on Paper 1 and 2 there will be one 6-, 8-, and 12-mark question, each on a different topic.

Each 12-mark question also has 3 extra marks for SPaG. So, they are actually worth 15 marks!

Aim to write your answer in fifteen minutes. There should be one sentence maximum for an introduction where you outline the case study and introduce any key concepts. Then dive straight into the question. Aim for two paragraphs if arguing for and against, otherwise three paragraphs (as you need a range of points), then a conclusion.

## How are 12-mark questions examined?

There are now four levels with an extra Level 4 requiring a 'comprehensive' answer.

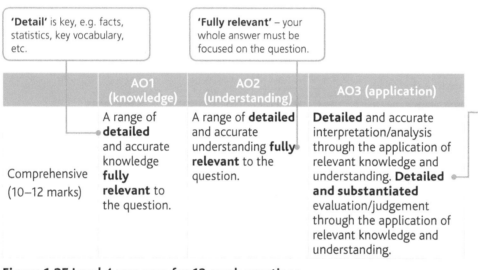

| | AO1 (knowledge) | AO2 (understanding) | AO3 (application) |
|---|---|---|---|
| Comprehensive (10–12 marks) | A range of **detailed** and accurate knowledge **fully relevant** to the question. | A range of **detailed** and accurate understanding **fully relevant** to the question. | **Detailed** and accurate interpretation/analysis through the application of relevant knowledge and understanding. **Detailed and substantiated** evaluation/judgement through the application of relevant knowledge and understanding. |

**'Detail'** is key, e.g. facts, statistics, key vocabulary, etc.

**'Fully relevant'** – your whole answer must be focused on the question.

**'Substantiated'** – back up your evaluation/judgements with evidence and explanation.

**Top tip:** Don't just look at the command word. There are often two or three different things required in the question to make it 'fully relevant' – so HUG your question!

**Figure 1.25 Level 4 response for 12-mark questions**

Typically, 4 marks are given to AO1 (knowledge), 4 marks to AO2 (understanding), and 4 marks to AO3 (analysis and evaluation). Consequently, the three key things required to do well on 12-mark questions are:

1 **Detail** – both knowledge (place-specific detail) and understanding (detail in your explanations)
2 **Evaluation**
3 **Structure**.

## Command words

Command words used in 12-mark questions are now fully evaluative.

| To what extent/how far do you agree? | Assess | Examine | Evaluate |
|---|---|---|---|

**Figure 1.26 Command words for 12-mark questions**

## Structuring your paragraphs – PEEL

Each paragraph needs structure. Therefore, like any banana, you should 'PEEL' each paragraph.

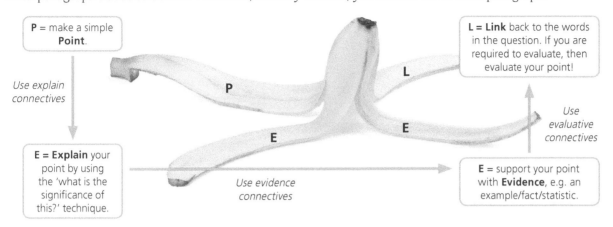

**P** = make a simple **Point**.

*Use explain connectives*

**E** = **Explain** your point by using the 'what is the significance of this?' technique.

*Use evidence connectives*

**E** = support your point with **Evidence**, e.g. an example/fact/statistic.

*Use evaluative connectives*

**L** = **Link** back to the words in the question. If you are required to evaluate, then evaluate your point!

**Figure 1.27 The PEEL technique for structuring a paragraph in an 8- or 12-mark question**

Let's now consider an example and think how to apply PEEL to it.

### Example 1

'*' = this means the answer will be marked on its SPaG and structure.

**1\*** **Evaluate** the <u>sustainability of the adaptations to a drought</u> caused by El Niño/La Niña in your <u>chosen case study area</u>.

**12 marks**

You need to make a judgement considering different factors, supported by evidence, on how sustainable each adaptation was. **AO3 analysis**

You must refer to a drought you have studied. **AO1 knowledge**

You must consider how each adaptation is sustainable/not sustainable. Aim for three adaptations, with a paragraph on each. **AO2 understanding**

For Example 1 you would PEEL three paragraphs (an adaptation in each paragraph), then write a mini conclusion.

Pick one adaptation and try using the PEEL structure yourself by writing in the boxes below. Aim to include:

- two explain connectives
- one evidence connective
- two evaluative connectives

**Top tip:** When evaluating a set of strategies, it is a good idea to compare across a spatial range, e.g. local/regional/national scale. You can also compare them by ranking them in your answer.

Adaptation: ................................................................................................................................

POINT:

........................................................................................................................

EXPLAIN:

........................................................................................................................

........................................................................................................................

........................................................................................................................

EVIDENCE:

........................................................................................................................

........................................................................................................................

LINK:

........................................................................................................................

........................................................................................................................

# Structuring an argument

For a 'to what extent/how far do you agree?' question, you would still use PEEL in each paragraph, but your overall structure would look a bit different, as you are arguing both for and against the statement.

## Example 2

2 'Humans have influenced river basins more than coastal landscapes.'
To what extent to do you agree?

(12 marks)

You need to explain the influence of humans on each of the two landscapes. **AO2 understanding**

You need to make a judgement considering the arguments for and against the statement, then justify your decision. **AO3 analysis**

Although it does not specifically mention the need for a case study, you have studied a case study for both landscapes and should refer to examples from them. **AO1 knowledge**

The structure to this question would have two main paragraphs and then a conclusion like this:

1. **Make your argument** that supports the statement. Use PEEL to structure your paragraph.

2. **Counter your argument** by opposing the statement. Use PEEL to structure your paragraph.

3. **Conclude** by weighing up the evidence on both sides and answer the question by coming to a judgement. Use the 'opinion line' to help you. Make sure you explain why you came to that judgement.

Figure 1.29 Constructing an argument for 'to what extent/how far do you agree?' questions

**Top tip for concluding:** Use the key words in the question in your conclusion to show you are answering the question.

1 Read the student answer on the next page.

Using everything you have learnt from pages 22–23, state four ways paragraph 1 is better than paragraph 2.

  i ........................................................................................................................

  ii ........................................................................................................................

  iii ........................................................................................................................

  iv ........................................................................................................................

2 Annotate around paragraph 2 different ways you could improve:
 - **P** (the Point) – can you develop it to make it clearer?
 - **E** (the Explanation) – could you add some explain connectives? How could you improve the explanation/add key geographical vocabulary?
 - **E** (the Evidence) – could you add evidence?
 - **L** (the Link back to the question) – could you add some evaluative connectives and phrases?

3 Write the conclusion using the opinion line and justify your decision.

4 Correct three SPaG mistakes in paragraph 2.

'Humans have influenced river basins more than coastal landscapes.' To what extent to do you agree?

**12 marks**

**Student answer:**

**Paragraph 1:**

Using the example of the river Tees drainage basin, it is clear humans have had a significant impact reducing flooding in the river basin. In the upper course, the building of Cow Green reservoir has controlled river flow and therefore reduced flooding downstream during peak flow after heavy rainfall. Down in the lower course the river was completely reshaped with 4km of the Mandale Loop being straightened. Whilst more recently dredging has increased the capacity of rivers and the building of the Tees Barrage has regulated river flow and prevented tidal flooding near the coast. Although flooding still occurs, humans have significantly modified the landscape throughout the whole drainage basin to reshape the river, control river flow and prevent flooding where possible. Due to the scale and extent of the impact flooding can have, this makes human influence more significant than in coastal landscapes.

**Paragraph 2:**

Humans effect coasts. Sea defenses can prevent erosion of cliffs, such as sea walls, Gabions and rock armour. Humans can also take sand from elsewhere and put more sand on the beach to protect it. Groynes prevent Longshore Drift so sand builds up. Some towns have built Promenades on the beachfront witch protects the land behind. Buildings on top of cliffs can increase pressure on the cliff leading to them collapsing. But Coastal defences are not everywhere and don't work all the time.

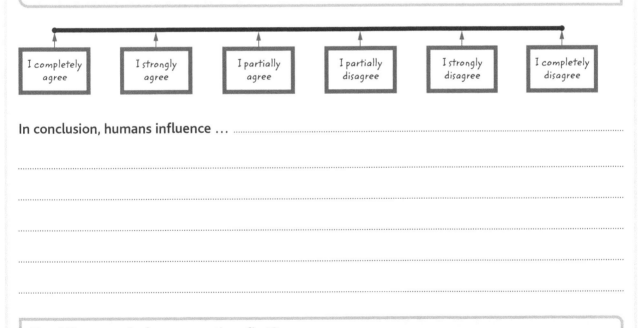

| I completely agree | I strongly agree | I partially agree | I partially disagree | I strongly disagree | I completely disagree |

In conclusion, humans influence …

.............................................................................................................................................

.............................................................................................................................................

.............................................................................................................................................

.............................................................................................................................................

**Should I answer the longer questions first?**

This is entirely up to you. They are certainly worth the most marks, so it is important not to run out of time for these questions. However, don't spend too long on them at the expense of other questions – be strict with yourself on timings!

# Chapter 2 Tackling Paper 1: Living in the UK Today

## Component 1.1: Landscapes of the UK

### Physical landscapes in the UK

The physical landscape of the UK contains upland, lowland and glaciated landscapes. They contain distinctive characteristics including their geology, climate, and types of human activity. **It is important to understand the key differences between the three landscapes.**

Complete the table below to summarise the differences between lowland, upland and glaciated landscapes. Use bullet points.

| | Lowland landscape | Upland landscape | Glaciated landscape |
|---|---|---|---|
| Definition | | • Elevated area possibly containing hills above 600m. | • Elevated area containing dramatic peaks and ridges. |
| Location | • Central and Southern England | | |
| Climate | | | |
| Geology | | | |
| Human activity | | | |

1. Which of the following characteristics is likely to be found in a glaciated landscape of the UK?

   A  Large cities

   B  Expansive areas for arable farming

   C  Regular freeze–thaw weathering due to cold climate

   D  Deposits of clays and gravel near the coast

   Write the correct letter in the box. ⬜  **1 mark**

2. Study **Figure 2.1** which shows a photograph of Campbeltown, Scotland. Identify two features of this lowland landscape.  **2 marks**

   1 ..................................................................................

   .....................................................................................

   2 ..................................................................................

   .....................................................................................

Figure 2.1 Campbeltown, Scotland.

**Top tips:**

1. Don't provide more than two features as you will only be marked on your first two!
2. Hint: What makes this landscape uniquely lowland?

# River landscapes (processes)

Geomorphic processes (processes that shape the Earth) shape distinctive river landscapes in the UK through weathering, mass movement, erosion, transportation and deposition. **Make sure you learn the names and definitions of the different types of each of these geomorphic processes!**

It is easy to confuse erosion and weathering. Write each process in the correct column in the table below.

Freeze–thaw    Hydraulic action    Oxidation    Abrasion

Biological    Attrition    Solution    Carbonation

| Erosion | Weathering |
|---|---|
|  |  |

**1** Define the process of saltation.    **1 mark**

............................................................................................................

............................................................................................................

> **Top tip:** Be as specific as you can – include both the cause of movement and how it moves.

**2** Define the process of abrasion.    **1 mark**

............................................................................................................

............................................................................................................

**3** Define the process of freeze–thaw weathering.    **1 mark**

............................................................................................................

............................................................................................................

**4** The table below names three weathering processes which take place within a river basin. Use arrows to match each process with the correct description.    **2 marks**

| Process | Description |
|---|---|
| Biological | The breakdown of rocks into smaller ones by water, ice or wind. |
| Mechanical | The disintegration of rocks caused by reactions. |
| Chemical | Rocks and river banks are broken down by living organisms, including plants and animals. |

# River landscapes (landforms)

Rivers create a range of distinctive landforms in the UK which include **waterfalls**, **gorges**, **V-shaped valleys**, **floodplains**, **levees**, **meanders** and **ox-bow lakes**.

1 **Explain the stages in the formation of a levee.** `3 marks`

### Key terms
Include the following key terms:
**deposition, floodplain, flood, river channel.**

**Top tip:** 3 marks = 3 stages to explain.

Showing 'stages' is often best done through drawing diagrams. For each stage in the formation of a levee, draw an annotated diagram below, using the vocabulary in the Key terms box to explain the stage.

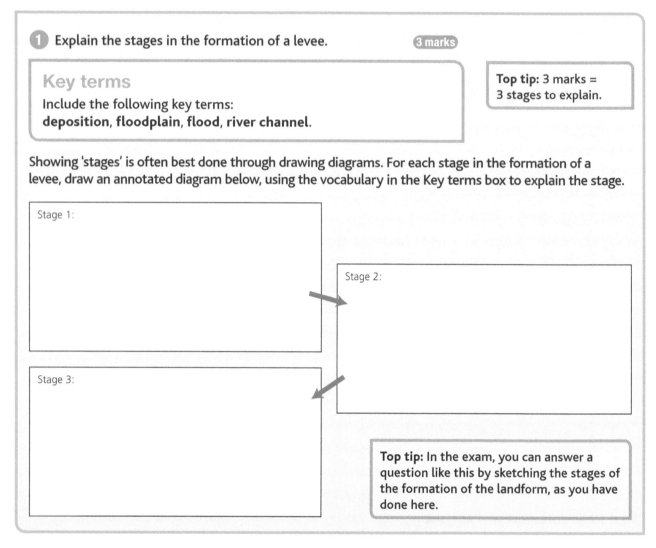

Stage 1:

Stage 2:

Stage 3:

**Top tip:** In the exam, you can answer a question like this by sketching the stages of the formation of the landform, as you have done here.

Now try explaining the stages of formation for all the river landscape landforms.

2 **Study Figure 2.2, which shows High Force, England. Label the features of this landform.** `3 marks`

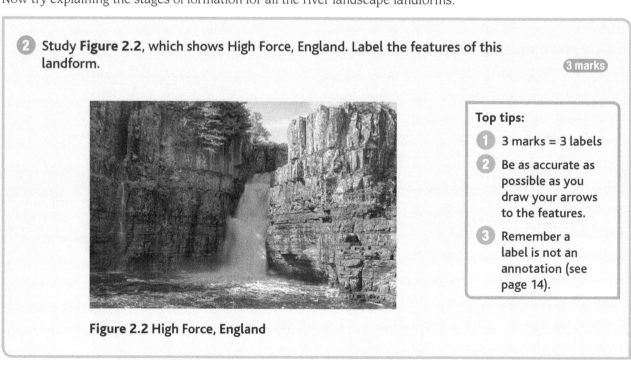

**Figure 2.2 High Force, England**

**Top tips:**

1 3 marks = 3 labels

2 Be as accurate as possible as you draw your arrows to the features.

3 Remember a label is not an annotation (see page 14).

# River landscapes (Case study 1)

You need to know one UK river basin in detail, with three focuses:

- how geomorphic processes operate at different scales and are influenced by geology and climate
- landforms created by these geomorphic processes
- the impact of human activity (including management) on geomorphic processes and the landscape.

**Name your chosen river basin in the UK:** ...............................................................

'Examine' means investigate/scrutinise the different influences. **AO3 analyse and evaluate**

Show you understand how both **geology** and **climate** influence geomorphic processes. Go back to page 27 if you need to remind yourself what geomorphic processes are. **AO2 understanding**

**Examine** the influence of geology and climate on geomorphic processes in your chosen river basin.

**8 marks**

Use examples from your river basin case study. **AO1 knowledge**

The focus of the question is on geology and climate, so structure it with two paragraphs. Use PEEL to help plan an answer to this question:

## 1 Geology

Point (outline your point)
.............................................................
.............................................................
.............................................................

↓

Explain (how geology influences processes)
.............................................................
.............................................................
.............................................................
.............................................................

↓

Evidence (using facts/examples)
.............................................................
.............................................................
.............................................................
.............................................................

↓

Link (examine the influence of geology on processes)
.............................................................
.............................................................
.............................................................
.............................................................

## 2 Climate

Point (outline your point)
.............................................................
.............................................................
.............................................................

↓

Explain (how geology influences processes)
.............................................................
.............................................................
.............................................................
.............................................................

↓

Evidence (using facts/examples)
.............................................................
.............................................................
.............................................................
.............................................................

↓

Link (examine the influence of geology on processes)
.............................................................
.............................................................
.............................................................
.............................................................

**How do I 'examine'?**

The following questions may help you 'examine' how geomorphic processes have been influenced:

- Sped up/slowed down?
- Change in location?
- Change of processes?
- Any processes become more/less important?

The key to accessing Level 3 'Thorough' (6–8 marks) for this question is to consider the **range of scales** the geomorphic processes are influenced on, from small-scale (specific landforms) to medium-scale (upper/middle/lower course) to large-scale (whole river basin).

# Coastal landscapes (processes)

Coastal landscapes are shaped by the same types of geomorphic processes as rivers. However, there are differences in the way they shape the land and some processes are more/less active. **Make sure you learn how geomorphic processes differ in coastal landscapes.**

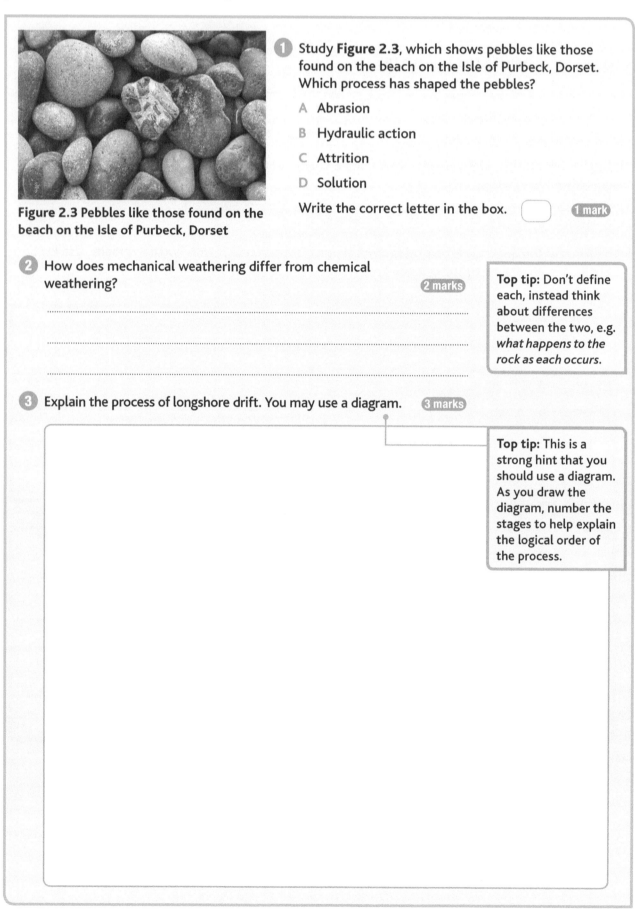

**Figure 2.3 Pebbles like those found on the beach on the Isle of Purbeck, Dorset**

**1** Study **Figure 2.3**, which shows pebbles like those found on the beach on the Isle of Purbeck, Dorset. Which process has shaped the pebbles?

A Abrasion

B Hydraulic action

C Attrition

D Solution

Write the correct letter in the box. ⬚ `1 mark`

**2** How does mechanical weathering differ from chemical weathering? `2 marks`

......................................................................................................

......................................................................................................

......................................................................................................

**Top tip:** Don't define each, instead think about differences between the two, e.g. *what happens to the rock as each occurs.*

**3** Explain the process of longshore drift. You may use a diagram. `3 marks`

**Top tip:** This is a strong hint that you should use a diagram. As you draw the diagram, number the stages to help explain the logical order of the process.

# Coastal landscapes (landforms)

Coasts create a range of distinctive landforms in the UK which include **headlands**, **bays**, **caves**, **arches**, **stacks**, **beaches** and **spits**.

1 a Explain how geomorphic processes influence the formation of an arch. You may use a diagram.

**4 marks**

**Top tip:** Don't just explain the formation of an arch. Instead focus on how geomorphic processes influence its formation. For each point you make, use the flow diagram below to help you.

**Geomorphic process**
↓
**Where it occurs**
↓
**What it does (its exact role in reshaping the landform)**

...................................................................................................................................

...................................................................................................................................

...................................................................................................................................

...................................................................................................................................

...................................................................................................................................

...................................................................................................................................

...................................................................................................................................

...................................................................................................................................

b Suggest how the arch might change in the future.

**2 marks**

...................................................................................................................................

...................................................................................................................................

...................................................................................................................................

...................................................................................................................................

## Want more?

For each of the other landforms pick one of these three different types of questions to answer:

1 Explain the stages in the formation of a headland/bay/cave/stack/spit/beach. **3 marks**

2 Explain how geomorphic processes influence the formation of a headland/bay/cave/stack/spit/beach. **4 marks**

3 Explain the formation of a headland/bay/cave/stack/spit/beach. **4 marks**

## Coastal landscapes (Case study 2)

Case study: ..................................................................................................

Use the table below to summarise your case study. Use bullet points to list key facts and statistics.

| | | |
|---|---|---|
| Geomorphic processes | | Top tip: Think about the different scales the geomorphic processes are operating on. E.g. individual beach or whole coastline? |
| Influences on geomorphic processes | On geology:<br><br>On climate: | |
| Landforms and features | | |
| Impact of human activity | On geomorphic processes (include a range of scales):<br><br>On the landscape: | |

---

Name your chosen coastal landscape in the UK: ....................................................................

Consider how humans and geomorphic processes influence the coastal landscape. Go back to page 30 if you need to remind yourself of coastal geomorphic processes. **AO2 understanding**

Use knowledge from your coastal landscape case study to exemplify influences. **AO1 knowledge**

**1** 'Humans have had a <u>more significant influence</u> on your <u>chosen coastal landscape</u> than <u>geomorphic processes</u>.' To what extent do you agree?

12 marks

Make a judgement by considering the arguments for and against. Use the 'opinion line' (see page 21). Then justify your decision. **AO3 analyse and evaluate**

Answer this question with:

- a one-sentence introduction
- two main paragraphs:
  - influence of humans
  - influence of geomorphic processes
- conclusion – *to what extent do you agree?*

The structure on page 33 will help you.

Top tip: If you have time, discuss two influences in each paragraph, so that you can cover a 'range' of influences to access Level 4.

Top tip: Be careful about what 'influences' you discuss here. The focus of the question is only on the landscape (e.g. processes and landforms).

**Introduction**

.............................................................................................................................

**Point:** On the one hand humans have ... ........................................................................

**Explain:** ..............................................................................................................

.............................................................................................................................

**Evidence:** ............................................................................................................

.............................................................................................................................

**Link:** Mark an 'X' on the line where you would judge the significance of human influences.

| No influence | Some influence | Very significant influence |
|---|---|---|

**Justification:** ......................................................

......................................................

> **Top tip:** Questions to consider as you justify:
> - How large an area has been affected?
> - How long has it affected/will it affect the area?
> - What is the rate of change?
> - Can it be managed?

**Point:** On the other hand, geomorphic processes have ... ........................................

**Explain:** ..............................................................................................................

.............................................................................................................................

**Evidence:** ............................................................................................................

.............................................................................................................................

**Link:** Mark an 'X' on the line where you would judge the significance of geomorphic influences.

| No influence | Some influence | Very significant influence |
|---|---|---|

**Justification:**

.............................................................................................................................

.............................................................................................................................

**Conclusion**

| Humans are the only influence | Humans had a larger influence | Both had the same influence | Geomorphic processes had a larger influence | Geomorphic processes are the only influence |
|---|---|---|---|---|

**Justification:**

.............................................................................................................................

.............................................................................................................................

## Trade in the UK

Through trade, the UK is connected to many countries and places around the world. **It is important to understand who the UK's current trading partners are and the UK's main exports and imports.**

**1** The table below shows three key terms associated with trade. Use arrows to match each term to the correct description. **2 marks**

| Key term | Description |
| --- | --- |
| Trade | Products brought into a country. |
| Import | Products taken out of a country. |
| Export | Involves the movement of goods and services across the world. |

**2** Study **Figure 2.4a** and **Figure 2.4b** showing the top ten UK imports and exports.

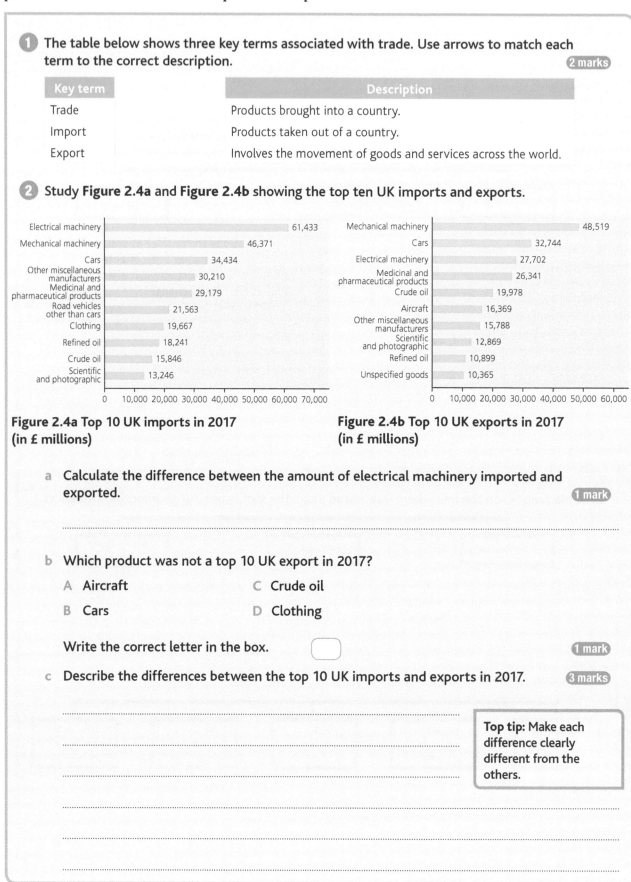

**Figure 2.4a Top 10 UK imports in 2017 (in £ millions)**

**Figure 2.4b Top 10 UK exports in 2017 (in £ millions)**

a Calculate the difference between the amount of electrical machinery imported and exported. **1 mark**

.............................................................................................................................

b Which product was not a top 10 UK export in 2017?

A Aircraft            C Crude oil

B Cars               D Clothing

Write the correct letter in the box. ⬭ **1 mark**

c Describe the differences between the top 10 UK imports and exports in 2017. **3 marks**

.............................................................................................................................

.............................................................................................................................

.............................................................................................................................

> **Top tip:** Make each difference clearly different from the others.

.............................................................................................................................

.............................................................................................................................

# Diversity in the UK

The UK is a diverse and unequal society. You need to be able to understand the UK's geographical patterns of diversity in **employment**, **average income**, **life expectancy**, **educational attainment**, **ethnicity** and **access to broadband**.

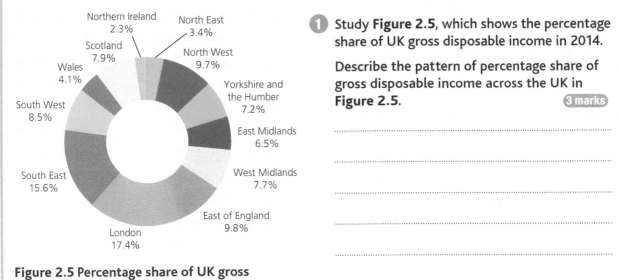

**Figure 2.5 Percentage share of UK gross disposable household income, 2014.**

1 Study **Figure 2.5**, which shows the percentage share of UK gross disposable income in 2014.

Describe the pattern of percentage share of gross disposable income across the UK in **Figure 2.5**. **3 marks**

....................................................................

....................................................................

....................................................................

....................................................................

....................................................................

**Top tip:** The temptation here is to list where there is high and low gross disposable income. Instead, aim to find an **overall pattern**.

2 Study **Figure 2.6**, which shows life expectancy at birth by sex and country, 2010–12.

Describe the pattern of life expectancy from birth for males and females. **4 marks**

**Top tip:** 1 Remember to refer to the figure!

2 Use 'GSD' to help you pick up the communication mark (see page 9).

| Country | Males | Females |
|---|---|---|
| England | 79.2 | 83.0 |
| Wales | 78.2 | 82.2 |
| Scotland | 76.6 | 80.8 |
| Northern Ireland | 77.8 | 82.3 |
| United Kingdom (average) | 78.9 | 82.7 |

**Figure 2.6 Life expectancy at birth by sex and country, 2010–12**

....................................................................

....................................................................

....................................................................

....................................................................

....................................................................

....................................................................

....................................................................

....................................................................

## Going further

Make sure you are also familiar with the UK spatial patterns of ethnicity, educational attainment and access to broadband.

# Development in the UK

The UK's development is not even. This is caused by **geographical location**, **economic change**, **infrastructure** and **government policy**.

**1** Study **Figure 2.5** (on page 35), which shows the percentage share of UK gross disposable income in 2014.

(on page 35)

Suggest how economic change may have caused regional variation in gross disposable income in **Figure 2.5**. **4 marks**

**Top tip:** Remember to refer to a Figure.

| Regional variation | Cause (because of economic change) |
|---|---|
| The north east, north west, and Wales have a very low combined share of gross disposable income (15.2%). | This is because … |
| | This is because … |

**2** Explain how government policy causes uneven development in the UK. **4 marks**

## Key terms

Use 'explain' connectives to help you develop your sentences in this question.

Pick two from: **So …, consequently …, as a result of …**

Then two of the following to develop the sentence even further: **In addition …, also …, furthermore …**

.......................................................................................................

.......................................................................................................

.......................................................................................................

.......................................................................................................

.......................................................................................................

.......................................................................................................

**Top tip:** An 'explain' question for 4 marks either requires two well-developed points, or four separate points. Aim to develop two points, and if you have time include a third point.

**3** Outline **one** way geographical location limits development in the UK.

Select one of the reasons below and explain how it limits development in the UK. **2 marks**

**Steep relief**                    **Distance from London**

**Distance from key transport routes**    **Distance from urban areas**

.......................................................................................................

.......................................................................................................

.......................................................................................................

.......................................................................................................

# Development in the UK (Case study 3)

For one UK area/region, you need to know the consequences of development, in particular what the consequences of economic growth and/or decline have been.

## CASE STUDY – UK place/region

Development has resulted in economic growth and/or decline in places/regions around the UK.

Name one place/region you have studied: ...................................................................................

No need to evaluate in this 8-mark question – just explain!

**Explain** the <u>economic</u> and <u>environmental consequences of development</u> in your <u>chosen place/region.</u>

**8 marks**

Use knowledge from your UK case study of a region that has seen economic growth and/or decline. **AO1 knowledge**

Explain how development has caused both **economic** *and* **environmental** consequences – this could include positives and negatives. One paragraph is needed for each. **AO2 understanding**

You only need to 'Explain', so you don't need to use 'L' in PEEL in structuring each paragraph. Instead just use PEE to answer this question.

| Economic consequences | Environmental consequences |
|---|---|
| **P**oint: ........................................ | **P**oint: ........................................ |
| **E**xplain: ........................................ | **E**xplain: ........................................ |
| **E**vidence: ........................................ | **E**vidence: ........................................ |

## Key terms

A key economic consequence of development is the **positive/negative multiplier effect**, which is the 'snowballing' of economic activity, either in a positive or negative direction, e.g. if new jobs are created, the newly employed have money to spend, which means more shop workers are needed. These shop workers pay taxes and spend their new money, creating yet more jobs in other industries.

Can you use the **multiplier effect** in your answer?

# Changing population in the UK

The UK's population is changing. You need to know:

- how the UK's population structure has changed since 1900 (including where the UK has been on the Demographic Transition Model)
- causes, effects and responses to an ageing population
- flows of immigration into the UK in the twenty-first century, and its social and economic impacts.

## UK population structure

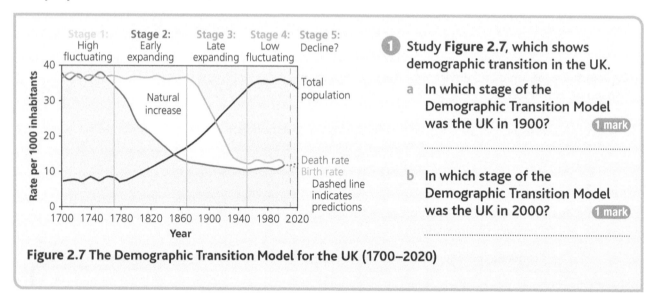

**1** Study **Figure 2.7**, which shows demographic transition in the UK.

   **a** In which stage of the Demographic Transition Model was the UK in 1900? `1 mark`

   **b** In which stage of the Demographic Transition Model was the UK in 2000? `1 mark`

Figure 2.7 The Demographic Transition Model for the UK (1700–2020)

## Ageing population

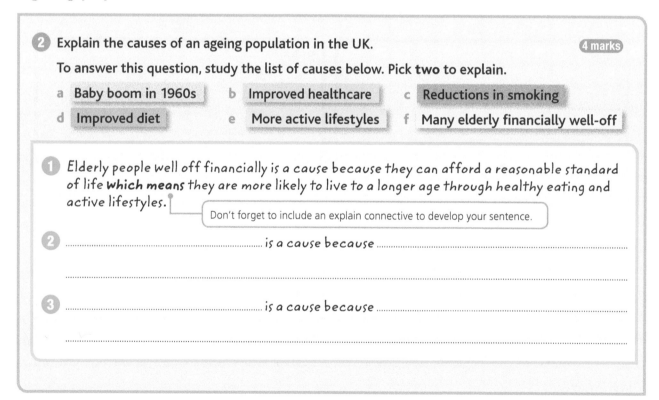

**2** Explain the causes of an ageing population in the UK. `4 marks`

To answer this question, study the list of causes below. Pick **two** to explain.

| a Baby boom in 1960s | b Improved healthcare | c Reductions in smoking |
| d Improved diet | e More active lifestyles | f Many elderly financially well-off |

**1** Elderly people well off financially is a cause because they can afford a reasonable standard of life **which means** they are more likely to live to a longer age through healthy eating and active lifestyles.

*Don't forget to include an explain connective to develop your sentence.*

**2** ............................................................ *is a cause because* ..............................................................

**3** ............................................................ *is a cause because* ..............................................................

## Want more?

**1** Explain the effects of/responses to an ageing population. `4 marks`

**2** Explain the two social/economic advantages/disadvantages of immigration into the UK. `4 marks`

# Urban trends in the UK

You need to know the causes and consequences of **suburbanisation**, **counter-urbanisation** and **re-urbanisation**.

1. The table below shows three urban trends in the UK. Use arrows to match each urban trend to the correct description.

   **2 marks**

| Urban trend | Description |
|---|---|
| Suburbanisation | Redevelopment of the inner urban areas, creating new homes and jobs and attracting people back to live in cities. |
| Counter-urbanisation | A movement of people from the inner city to the outskirts. |
| Re-urbanisation | A movement of people from urban areas to rural areas. |

**CB1 Development in Cambridge**

- £725 million redevelopment scheme.
- 1,000 student apartments, 350 homes built.
- Now contains green spaces, cafes, restaurants, shops and a Microsoft office.

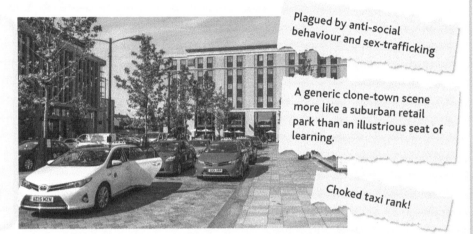

Plagued by anti-social behaviour and sex-trafficking

A generic clone-town scene more like a suburban retail park than an illustrious seat of learning.

Choked taxi rank!

**Figure 2.8 An inner-city development scheme in Cambridge and its impacts**

2. Study **Figure 2.8**, which shows the impacts of an inner-city development scheme in Cambridge.

   a Which urban trend is this an example of?

   **1 mark**

   .........................................................................................................................

   b Using **Figure 2.8**, discuss the consequences of this urban trend in the UK.

   **4 marks**

   Use the structure provided in the table below to write an answer to Question 2b.

| Consequence | Evidence connective | Evidence from Figure 2.8 | Explain connective | Explanation of consequence |
|---|---|---|---|---|
| Rejuvenation of inner city ... | ...evidenced by... | ... Figure 2.8 saying there has been the arrival of new cafes, restaurants, shops and a Microsoft Office... | ... as a result... | ...there are new job opportunities resulting in the multiplier effect, with increasing amounts of money spent in the local area. |

**Want more?**

1. Explain the causes of suburbanisation/counter-urbanisation/re-urbanisation.

   **4 marks**

**Case study:** ........................................................................................................

Use the table below to summarise your case study. Use bullet points to list key facts and statistics. Be as specific as possible.

| | | |
|---|---|---|
| **Influence of the city on ...** | Region: | |
| | Country: | |
| | Wider world: | |
| **Influence of migration on the city's ...** | Growth: | |
| | Character: | |
| **Influence on ways of life in the city ...** | Culture: | |
| | Ethnicity: | |
| | Housing: | |
| | Leisure: | |
| | Consumption: | |
| **Contemporary challenges ...** | Housing availability: | |
| | Transport provision: | |
| | Waste management: | |
| **Sustainable strategies to overcome one challenge ...** | | |

**Top tip:** You only need to look at the management of **one** challenge.

## CASE STUDY – a major city in the UK

**Name of UK city:** .................................................................................................................................

Contemporary challenges affecting UK cities include housing availability, transport provision and waste management.

> Use knowledge from your UK city case study. **AO1 knowledge**

Underline: **Evaluate** the <u>sustainability of strategies</u> used in your <u>chosen city</u> to overcome **one** challenge. `12 marks`

> Make a judgement on each strategy considering different factors and using evidence. **AO3 analysis and evaluation**

> Consider how sustainable different strategies are to overcome one urban challenge in your chosen city. Aim for three or four strategies, with a paragraph for each. **AO2 understanding**

To help answer this question, use the radar diagram below.

1. In the box, briefly explain how each strategy is/is not economically, socially and environmentally sustainable. Use facts and statistics as evidence.

2. Mark an X on each line in the middle for how sustainable you think each strategy is overall.

3. Join up your four 'Xs' to make a radar diagram, to help you write a conclusion.

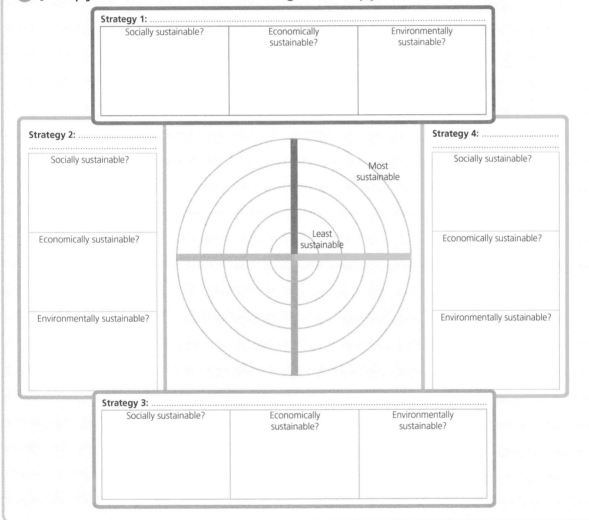

**Strategy 1:** ....................................................

| Socially sustainable? | Economically sustainable? | Environmentally sustainable? |
|---|---|---|
| | | |

**Strategy 2:** ....................................................

- Socially sustainable?
- Economically sustainable?
- Environmentally sustainable?

**Strategy 4:** ....................................................

- Socially sustainable?
- Economically sustainable?
- Environmentally sustainable?

Most sustainable / Least sustainable

**Strategy 3:** ....................................................

| Socially sustainable? | Economically sustainable? | Environmentally sustainable? |
|---|---|---|
| | | |

## Key terms

Something is **sustainable** if it meets the current generation's economic, social and environmental needs, without compromising the needs of future generations. This can be illustrated by the sustainability stool, which needs all three legs to keep it standing.

## Want more?

1. 'Transport provision is the greatest contemporary challenge affecting your chosen city.' How far do you agree? `12 marks`

# Component 1.3: UK Environmental Challenges

## Extreme weather

The UK's climate is unique for its latitude, which can create extreme weather conditions. Three factors affect the weather in the UK: **air masses** (make sure you know what weather the five air masses bring), the **North Atlantic Drift**, and **continentality**.

Air masses cause extreme weather conditions in the UK, including extremes of wind, temperature and precipitation. **It would be helpful for you to know an example of each in the UK**.

N     0            400 km

Nottingham

Plymouth

**Figure 2.9 Air mass (red arrow) influencing the UK**

---

1. Which air mass brings very cold conditions in the winter with snow, but very hot sunny weather in the summer?

   A  Polar Continental    C  Polar Maritime        Write the correct letter in the box.  ☐    **1 mark**

   B  Arctic Maritime      D  Tropical Maritime

2. Study **Figure 2.9**, which shows an air mass (the red arrow) influencing the UK.

   a  Describe the weather Plymouth would expect to receive from this air mass.    **2 marks**

   ........................................................................................................................

   ........................................................................................................................

   ........................................................................................................................

   > **Top tip:** 2 marks = 2 separate comparisons. Opposite points count as the same point!

   b  Compare how continentality influences the weather in Nottingham and Plymouth.    **2 marks**

   ........................................................................................................................

   ........................................................................................................................

   ........................................................................................................................

3. Explain how air masses can bring extreme temperatures/winds in the UK.    **4 marks**

   Use the table below to plan an answer to both versions of Question 3.

   |  | Extreme temperatures | Extreme winds |
   | --- | --- | --- |
   | Air mass |  |  |
   | What weather does it bring? |  |  |
   | How does it bring it? |  |  |
   | Examples |  |  |

# Extreme weather (Case study 5)

You need to know one UK flood event caused by extreme weather conditions including:

- causes (including the role of extreme weather conditions)
- effects on the people and the environment
- management at a variety of scales.

**UK flood event:** ..................................................................................................................................................

**Use the table below to summarise this case study. Use bullet points to list key facts and statistics.**

| Key facts | Date:<br>Precipitation:<br>Other key facts: | | Location (draw a sketch map): |
|---|---|---|---|
| Causes | Physical causes (including how caused by extreme weather):<br><br>Human causes: | | |
| Impacts | Social:<br><br>Economic:<br><br>Environmental: | | |
| Management | Spatial scale<br>Individual | Short-term response | Long-term response |
| | Local | | |
| | Whole river basin | | |

**Top tip:** By considering both spatial (space) and temporal (time) scales, you are considering the variety of scales that the specification asks for.

## CASE STUDY – UK flood event caused by extreme weather conditions

**Name of UK flood event:** ...................................................................................................................

**Assess** the <u>management</u> of the flood event at **two** <u>different scales</u>.    **6 marks**

| | |
|---|---|
| Offer a reasoned judgement informed by relevant facts of the management of the flood. **AO3 analysis and evaluation** | Consider how the flood was managed on **two** scales (either temporal or spatial). A paragraph for each scale. **AO2 understanding** |

Read the two example answers below. They are different levels, but neither got full marks. Use the activities by the table to help you.

    a  **Mark** each answer by using the marking grid below each one. Make a judgement on whether each of the criteria is Basic, Reasonable, or Thorough.

    b  **Improve** the answers by writing your own answer to the question.

### Student answer 1:

In 2014 the UK had the worst storms for twenty years, with Somerset having more than 65km2 flooded. After the 2014 Somerset floods, the Prime Minister announced, 'money was no object' and as a result the government responded in lots of ways.

They dredged the River Tone and River Parrett. The Environment Agency brought 62 pumps all the way from the Netherlands. In March 2014, a twenty-year flood action plan for Somerset was created. The village of Muchelney had their road raised. Also, lots of defences were repaired such as embankments.

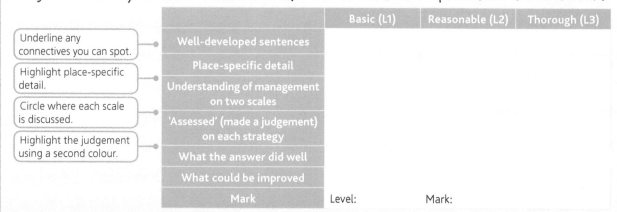

| | Basic (L1) | Reasonable (L2) | Thorough (L3) |
|---|---|---|---|
| Well-developed sentences | | | |
| Place-specific detail | | | |
| Understanding of management on two scales | | | |
| 'Assessed' (made a judgement) on each strategy | | | |
| What the answer did well | | | |
| What could be improved | | | |
| Mark | Level: | | Mark: |

- Underline any connectives you can spot.
- Highlight place-specific detail.
- Circle where each scale is discussed.
- Highlight the judgement using a second colour.

### Student answer 2:

After the Somerset Levels flood in early 2014, there were a number of short-term responses. Soldiers helped residents who had been cut off by the floodwater, moreover lots of pumps were used to reduce the floodwater levels in key locations. Whilst these did not stop the flood occurring, they sought to reduce the impacts to people.

There were also responses over a longer timescale too. Across the county defences were repaired to reduce the likelihood of a future flood happening again. Plans were also set in place to dredge the rivers, so that they had a greater storage capacity. Roads would be raised too, because that would reduce the impact on transport infrastructure when the next flood occurred, and prevent residents being cut off by the floodwater. These longer-term responses hope to reduce the likelihood of a future flood of the same scale.

| | Basic (L1) | Reasonable (L2) | Thorough (L3) |
|---|---|---|---|
| Well-developed sentences | | | |
| Place-specific detail | | | |
| Understanding of management on two scales | | | |
| 'Assessed' each strategy | | | |
| What the answer did well | | | |
| What could be improved | | | |
| Mark | Level: | | Mark: |

# Ecosystems

In the UK, humans use, modify and change ecosystems and environments to obtain food, energy and water. The ways they do this include:

- mechanisation of farming
- commercial fishing to provide food
- wind farms and fracking to provide energy
- reservoirs and water transfer schemes to provide water.

**Figure 2.10 A wind farm in the UK**

**1** Study **Figure 2.10**, which shows a wind farm in the UK.

**Tick three impacts wind farms have on the environment.** `2 marks`

| Impact | |
|---|---|
| Wind farms cheap to produce compared to other sources of energy. | ☐ |
| The wind turbines create a lot of noise. | ☐ |
| Greenhouse gas emissions are produced by the manufacturing and transport of turbines to the site. | ☐ |
| Wind farms are found on land and at sea in the UK. | ☐ |
| Onshore wind farms found on high ground where the wind is highest. | ☐ |
| Many consider wind turbines to be an eyesore, disturbing the natural views of the landscape. | ☐ |

**2** Explain how the mechanisation of farming/commercial fishing/reservoirs and water transfer schemes affect ecosystems.

Match up the point with the correct explanation and further explanation for the different versions of Question 2. `4 marks`

**Top tip:** Watch out for the wording 'ecosystems' and the 'environment'! **Environment** = surrounding area. **Ecosystem** = the interaction between the environment and the organisms living in it.

| Point | Explanation | Further explanation |
|---|---|---|
| To house machinery and produce food on a large-scale, farms have increased in size ... | ... consequently, there is the accidental death of unplanned species ... | ... therefore, there are not enough young fish left in the sea to breed and replace those that have been lost. |
| Use of chemical fertilisers and pesticides ... | ... as a result, river channels silt up and there is an increase in salinity ... | ... which affects aquatic plant and animal life. |
| Over-fishing of popular fish species such as cod ... | ... as a result, hedgerows have been destroyed ... | ... causing algae blooms, which starve other plants of oxygen, and they eventually die. |
| Use of large nets in commercial fishing catches other unplanned species ... | ... so too many young fish are caught ... | ... resulting in changes in the number of species in their food webs, which impacts other species in the food web. |
| Building of reservoirs limits the flow of water downstream ... | ... by leading to nutrient imbalances ... | ... which affects the habitats of small mammals living there and the food webs they are in. |
| Transfer of water from one region to another impacts the local ecology ... | ... results in the run-off of these into local rivers and lakes and the occurrence of eutrophication ... | ... causing fish species to migrate from their natural habitats. |

# Energy sources

The UK uses a range of renewable and non-renewable energy sources.

**1** Describe the difference between renewable and non-renewable energy sources. `1 mark`

.................................................................................................................................

.................................................................................................................................

**2** Study **Figure 2.11**, which shows the sources from which electricity is generated in the UK in 2016 and 2017.

|  | 2016 | 2017 |
|---|---|---|
| Coal | 9.0% | 6.7% |
| Gas | 42.2% | 39.7% |
| Oil and Other | 3.1% | 3.2% |
| Nuclear | 21.1% | 20.9% |
| Renewables | 22.0% | 29.4% |

**Figure 2.11 Sources from which electricity is generated in the UK in 2016 and 2017**

a Calculate the percentage change in electricity generated from renewables between 2016 and 2017. `2 marks`

.................................................................................................................................

.................................................................................................................................

b Describe how the energy mix for electricity generated has changed between 2016 and 2017. `3 marks`

.................................................................................................

.................................................................................................

.................................................................................................

> **Top tip:** Look for patterns, e.g. renewables versus non-renewables.

**3** a Match the **eight** energy sources to the correct definition below. **Two** energy sources are not needed.

Coal    Hydro    Biomass    Geothermal    Solar    Wind    Gas    Oil    Nuclear    Tidal

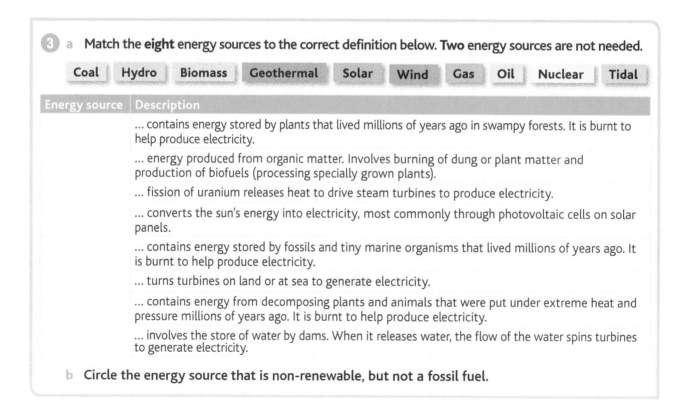

| Energy source | Description |
|---|---|
|  | ... contains energy stored by plants that lived millions of years ago in swampy forests. It is burnt to help produce electricity. |
|  | ... energy produced from organic matter. Involves burning of dung or plant matter and production of biofuels (processing specially grown plants). |
|  | ... fission of uranium releases heat to drive steam turbines to produce electricity. |
|  | ... converts the sun's energy into electricity, most commonly through photovoltaic cells on solar panels. |
|  | ... contains energy stored by fossils and tiny marine organisms that lived millions of years ago. It is burnt to help produce electricity. |
|  | ... turns turbines on land or at sea to generate electricity. |
|  | ... contains energy from decomposing plants and animals that were put under extreme heat and pressure millions of years ago. It is burnt to help produce electricity. |
|  | ... involves the store of water by dams. When it releases water, the flow of the water spins turbines to generate electricity. |

b Circle the energy source that is non-renewable, but not a fossil fuel.

4. Study **Figures 2.12a** and **2.12b**. **Figure 2.12a** shows the renewable energy sources from which electricity was generated in the UK from 2007 to 2017. **Figure 2.12b** shows the percentage of each renewable energy source from which electricity was generated in 2017.

a In which year did the renewable energy total, from which electricity was generated, **(1 mark)** decrease?

.......................................................................................................................................................

b Complete the percentage each energy source contributes to electricity generation **(2 marks)** in **Figure 2.12b** below.

c Describe the trends for renewables' contribution to electricity generation between 2007 and 2017. **(3 marks)**

.......................................................................................................................................................

.......................................................................................................................................................

.......................................................................................................................................................

.......................................................................................................................................................

.......................................................................................................................................................

**Top tip:** When describing a 'trend', 'chunk' up the different sections and use good descriptive adjectives.

Watch out – this is the percentage of **all** energy sources, not just renewables! Divide the amount generated by that source, by the renewables total. Then multiple it by 29.4 (instead of 100).

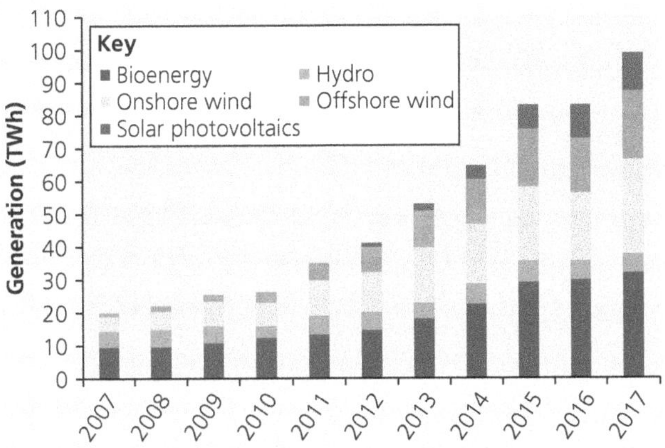

Figure 2.12a Renewable energy sources from which electricity was generated in the UK from 2007 to 2017

**Key**
- Bioenergy
- Hydro
- Onshore wind
- Offshore wind
- Solar photovoltaics

| | Generation (TWh) | Percentage contribution from which electricity is generated |
|---|---|---|
| Onshore wind | 28.7 | 8.5% |
| Offshore wind | 20.9 | |
| Hydro | 5.9 | 1.8% |
| Solar photovoltaics | 11.5 | |
| Bioenergy | 31.8 | 9.5% |
| All renewables | 98.9 | 29.4% |

Figure 2.12b The contribution of each renewable energy source to electricity generation in the UK in 2017

Give an account that addresses a range of ideas on the sustainability of the UK's energy supply. **AO3 analysis and evaluation**

Consider how sustainable the UK's energy supply is – remember the sustainability stool! **AO2 understanding**

Use your own knowledge about the UK's energy supply. **Figures 2.11** and **2.12a** might help! **AO1 knowledge**

5. **Discuss** the sustainability of the UK's energy supply. **(6 marks)**

There are two ways you could structure a response to this question:
- Write two paragraphs using the 'argument for and counter-argument' technique (see page 24).
- Write three mini-paragraphs considering how the UK's energy supply is economically, socially, and environmentally sustainable.

# Energy management

You need to know:

- how patterns of energy supply and demand have changed since 1950, and how government decision making and international organisations have influenced these
- how successful strategies for sustainable use and management of energy have been, at local and national scales
- how renewable energy has developed, and its impacts on people and the environment
- the extent to which non-renewable energy could and should contribute to the UK's future energy supply
- economic, political and environmental factors affecting UK energy supply in the future.

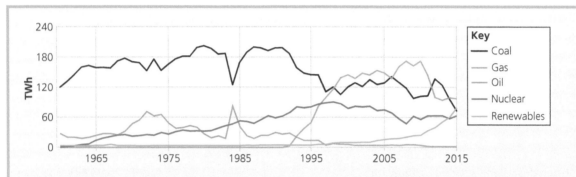

**Figure 2.13 Trends in UK electricity generation by source, from 1960 to 2015**

**1** Study **Figure 2.13**, which shows the UK electricity generation by source from 1960 to 2015.

> **Top tip:** Aim for four suggestions.

Using **Figure 2.13**, suggest how government decision making has influenced energy supply since 1960.

> **Top tip:** Make a specific reference to a change on the graph in **Figure 2.13**.

Use the table to help structure an answer to Question 1. The first row has been done for you. Write **two** more.     **4 marks**

| Observed change in energy supply | Government decision |
|---|---|
| Sharp drop in the early 1980s in electricity supplied by coal ... | ... caused by the government choosing to close multiple coal mines in the early 1980s. |
| | |
| | |

> Offer a reasoned judgement informed by relevant facts on the success of the sustainable management of energy. **AO3 analysis and evaluation**

> Consider how successful the sustainable management of energy has been on a local/national scale. Remember the sustainability stool! **AO2 understanding**

**2** Assess the success of the sustainable management of energy on a local/national scale.     **6 marks**

> Use knowledge of a local/national scale example. **AO1 knowledge**

Like Question 5 on page 47, this could be structured by either of the suggested ways. Let's try the second. Mark an 'X' on each line and then justify your decision.

Top tip: Remember specific examples and facts to help you explain your decisions!

| Example | Local scale | National scale |
|---|---|---|
| Economically sustainable? | Very ———————————————— Not<br>Why? | Very ———————————————— Not<br>Why? |
| Socially sustainable? | Very ———————————————— Not<br>Why? | Very ———————————————— Not<br>Why? |
| Environmentally sustainable? | Very ———————————————— Not<br>Why? | Very ———————————————— Not<br>Why? |
| Overall, how successful? | Very ———————————————— Not<br>Why? | Very ———————————————— Not<br>Why? |

3 Explain how economic/political/environmental factors will affect UK energy supply in the future.

**4 marks**

Read the different factors below. For each type of factor use a different coloured highlighter to differentiate whether it is an economic, political or environmental factor.

High cost of building new nuclear power stations and decommissioning old ones.

Many are concerned about the radioactive waste produced by nuclear power and the dangers of a radioactive leak.

Will the government continue to encourage the renewable sector with grants and subsidies?

Different political parties may adopt different stances on fracking.

North Sea supplies of oil and gas are declining, making it increasingly expensive to extract there.

Fracking pollutes groundwater aquifers, and small earthquakes have been felt nearby.

UK Climate Change Act commits the UK to reduce greenhouse gas emissions by 80 per cent by 2050.

There is a high cost of constructing renewable energy alternatives such as wind farms, tidal barrages and HEP.

Energy security needs to be ensured by importing natural gas from a range of politically stable countries.

Key

Economic factor

Political factor

Environmental factor

**Pick two factors to answer one variation of Question 3. Remember to use explain connectives!**

4 **To what extent** could and should non-renewable energy contribute to the UK's future energy supply?

**12 marks**

Make a judgement considering the arguments for and against 'could' and 'should'. To access Level 4 it is crucial you make a judgement on both and justify your decision. **AO3 analysis and evaluation**

Consider the role of non-renewable energy – could it *and* should it contribute to the UK's future energy supply? One paragraph on each. **AO2 understanding**

Use knowledge of the current UK energy mix to help answer 'could' and 'should'. **AO1 knowledge**

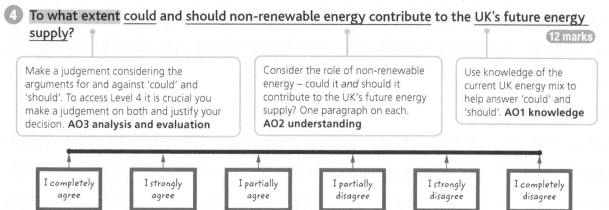

I completely agree — I strongly agree — I partially agree — I partially disagree — I strongly disagree — I completely disagree

To help you plan an answer to this question, mark two 'Xs' mark on the 'opinion line' for your response to 'could' and 'should' the UK use non-renewables in its future energy supply.

Now answer this question on a separate piece of paper.

## Want more?

1 Examine the impacts of one type of renewable energy on people and the environment.

**8 marks**

# Chapter 3 Tackling Paper 2: The World Around Us

## Component 2.1: Ecosystems of the Planet

### Global ecosystems

You need to know the global distribution, climate, and the plants and animals that operate within the following ecosystems: **polar regions**, **coral reefs**, **grasslands (temperate and tropical)**, **temperate forests**, **tropical rainforests** and **hot deserts**.

1. Match the **seven** key terms to the correct definition below. **One** is not needed.

| Ecosystem | Biome | Biotic | Flora | Abiotic | Climate | Fauna | Interdependence |

| Key term | Description |
|---|---|
| | ... large-scale ecosystems that are spread across continents and have plants and animals unique to them. |
| | ... the physical, non-living parts of the ecosystem. |
| | ... the reliance of every form of life on other living things and on the natural resources in its environment. |
| | ... animals in an ecosystem. |
| | ... all the living elements of the ecosystem. |
| | ... plants in an ecosystem. |
| | ... a community of living organisms interacting with the non-living parts of the environment. |

2. To help identify the differences between each ecosystem, copy the table below onto A3 paper, and summarise the key characteristics of each ecosystem. Leave the tropical rainforests and coral reefs rows for now – you will complete them later in this chapter. Your completed table can be used to answer the questions on pages 52 and 55.

| | Location | Climatic features | Flora (two examples and details, e.g. adaptations) | Fauna (two examples and details, e.g. adaptations) |
|---|---|---|---|---|
| Polar regions | | | | |
| Temperate forest | | | | |
| Temperate grassland | | | | |
| Tropical grassland | | | | |
| Hot deserts | | | | |
| Tropical rainforests | | | | |
| Coral reefs | | | | |

3. Study the List below of components found in an ecosystem.
   a. Circle the three biotic components..

| Wind | Soil | Plants | Nutrients | Humans | Precipitation | Fungi | Sunlight |

# The global distribution of ecosystems

It is likely that exam questions will ask you to 'describe the distribution' and 'identify' ecosystems on a map. Make sure you can accurately describe the location of each ecosystem.

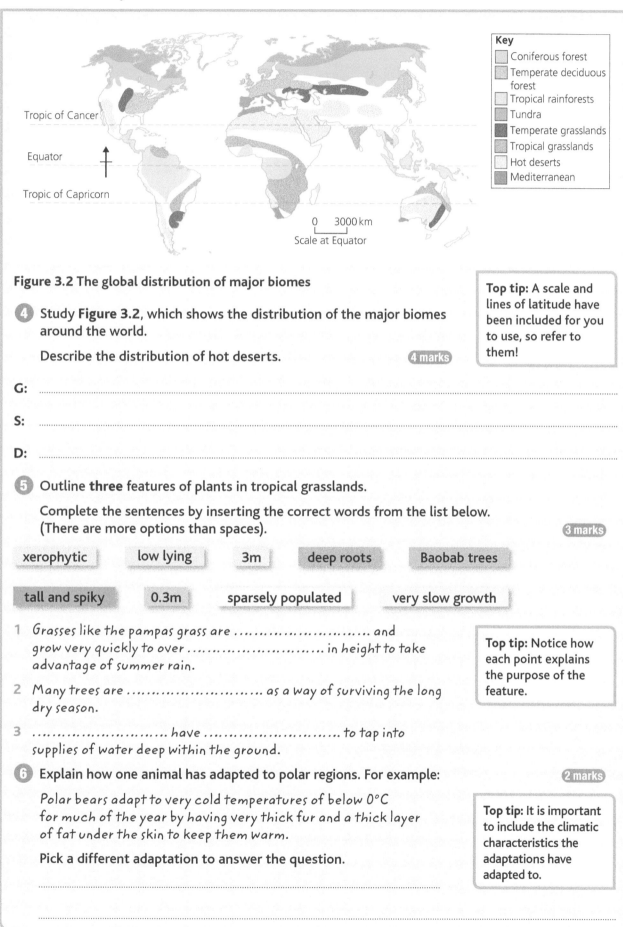

**Figure 3.2 The global distribution of major biomes**

**4** Study **Figure 3.2**, which shows the distribution of the major biomes around the world.

Describe the distribution of hot deserts. **4 marks**

> **Top tip:** A scale and lines of latitude have been included for you to use, so refer to them!

**G:** ...........................................................................................................

**S:** ...........................................................................................................

**D:** ...........................................................................................................

**5** Outline **three** features of plants in tropical grasslands.

Complete the sentences by inserting the correct words from the list below. (There are more options than spaces). **3 marks**

| xerophytic | low lying | 3m | deep roots | Baobab trees |

| tall and spiky | 0.3m | sparsely populated | very slow growth |

1 Grasses like the pampas grass are ........................... and grow very quickly to over ........................... in height to take advantage of summer rain.

> **Top tip:** Notice how each point explains the purpose of the feature.

2 Many trees are ........................... as a way of surviving the long dry season.

3 ........................... have ........................... to tap into supplies of water deep within the ground.

**6** Explain how one animal has adapted to polar regions. For example: **2 marks**

Polar bears adapt to very cold temperatures of below 0°C for much of the year by having very thick fur and a thick layer of fat under the skin to keep them warm.

**Pick a different adaptation to answer the question.**

> **Top tip:** It is important to include the climatic characteristics the adaptations have adapted to.

.................................................................................................................

.................................................................................................................

# Tropical rainforests

For tropical rainforests, you also need to know the water and nutrient cycle in tropical rainforests and the location of the following rainforests: **Amazon**, **Central American**, **Congo River Basin**, **Madagascan**, **South East Asian** and **Australasian**.

Complete the tropical rainforests row on the table you started on page 50.

| | Location | Climatic features | Flora (two examples and details, e.g. adaptations) | Fauna (two examples and details, e.g. adaptations) |
|---|---|---|---|---|
| Tropical rainforests | | | | |

1 Study **Figure 3.3** which shows the global distribution of tropical rainforests.

Identify the names of the rainforests marked A, B and C.

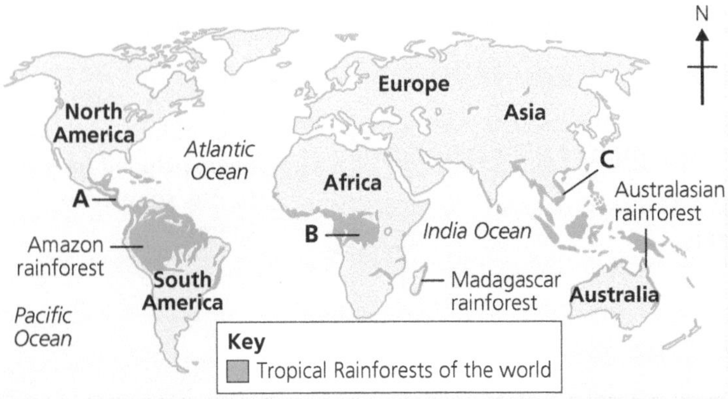

A: ..............................................................

B: ..............................................................

C: ..............................................................

Figure 3.3 The global distribution of tropical rainforests

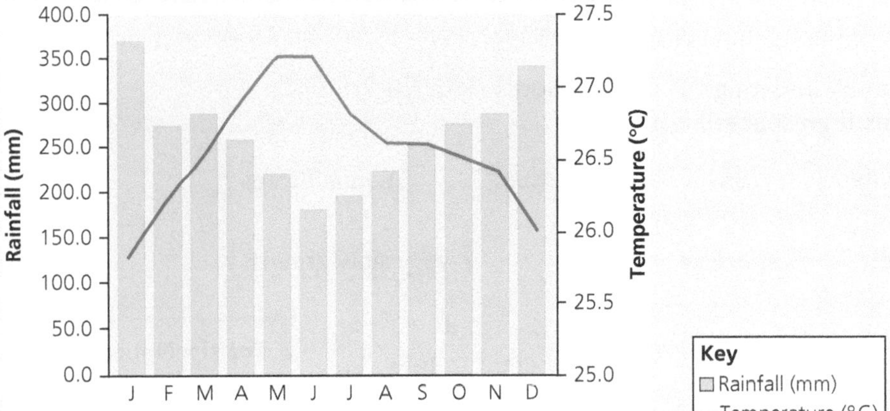

Figure 3.4 Climate graph for Sibu, Malaysia

**Top tip:** Rainfall is always shown as blue bars on the left-hand y-axis. Temperature is the red line and right-hand y-axis.

**Top tip:** Use the framework below to guide any description of a climate graph. Remember to provide specific data of the climate.

2 Study **Figure 3.4**, which shows the climate for Sibu, Malaysia in the South-East Asian rainforest.

Describe the climate for Sibu, Malaysia. **4 marks**

Describe the rainfall pattern: ..........................................................................................................................

..................................................................................................................................................................

Describe the temperature pattern: ................................................................................................................

..................................................................................................................................................................

How do the temperature and rainfall link together? ....................................................................................

..................................................................................................................................................................

**3** Explain how water is recycled in the tropical rainforest. Use a flow diagram and the Key terms to answer Question 3.    **4 marks**

**Top tip:** You will need four separate points in your flow diagram.

## Tropical rainforests (Case study 6)

**Tropical rainforest case study:** ......................................................................................................................

Use the table below to summarise this case study. Use bullet points to list key facts and statistics. Be as specific as possible.

| | |
|---|---|
| **Interdependence** | Climate: |
| | Soil: |
| | Water: |
| | Plants: |
| | Animals: |
| | Humans: |
| **Value** | To humans: |
| | To the planet: |
| **Threats to biodiversity** | |
| **Sustainable use and management** | |

**Top tip:** Aim for a range of spatial scales to help you to evaluate in the 8- and 12-mark questions.

# CASE STUDY – a tropical rainforest.

**Name of tropical rainforest:** ......................................................................................................

Consider the idea of interdependence of biotic (humans, plants and animals) *and* abiotic (climate, soil and water) components.
**AO2 understanding**

**4** **Explain** the <u>interdependence of abiotic</u> and <u>biotic components</u> in the tropical rainforest you have studied. **8 marks**

Use knowledge from your tropical rainforest. **AO1 knowledge**

Use 'PEE' to structure this answer by using the table below to plan your answer. An example has been done for you.

| POINT: Abiotic/biotic interdependence | EXPLAIN: | Connective to explain further | Further explanation (the significance of the interdependence) | EVIDENCE: |
|---|---|---|---|---|
| 1. Biotic components can be dependent on abiotic components. | The warm and wet climate (an abiotic factor) ensures annual precipitation is high, and the nutrients are rapidly recycled. | ...consequently... | ...the conditions are ideal for fast growth of vegetation and the maintaining of the rainforest. | Average temperature is 28°C and annual rainfall is over 2500mm. |
| 2. Abiotic components can be dependent on biotic components. | → | → | → | |
| 3. Biotic components can be dependent on biotic components. | → | → | → | |
| 4. Abiotic components can be dependent on abiotic components. | → | → | → | |

## Want more?

**1** Assess the statement 'tropical rainforests are of more value to humans than the planet'. **8 marks**

**2** 'Tropical rainforests are managed sustainably.' Using the tropical rainforest you have studied, how far do you agree? **12 marks**

# Coral reefs

As with tropical rainforests, you need to know global distribution, climate, and the plants and animals that operate within coral reefs. You also need to know the **nutrient cycling process** in coral reefs and the location of the following coral reefs: **Great Barrier Reef**, **Red Sea Coral Reef**, **New Caledonia Barrier Reef**, **the Mesoamerican Barrier Reef**, **Florida Reef** and **Andros Coral Reef**.

Complete the coral reefs row on the table you started on page 50. The 'climatic features' is less relevant for coral reefs; instead think of 'environmental conditions' (there are **three** key ones).

| | Location | Climatic features | Flora (two examples and details, e.g. adaptations) | Fauna (two examples and details, e.g. adaptations) |
|---|---|---|---|---|
| Coral reefs | | | | |

1. State **two** warm-water coral reefs that can be found near Central America. **2 marks**

   1. ...................................................................................................................

   2. ...................................................................................................................

2. Outline **two** conditions needed for coral reefs to form. **2 marks**

   .........................................................................................................................

   .........................................................................................................................

   **Top tip:** Include specific details where you can to secure the mark for your point.

3. The table below shows four features of the coral reef nutrient cycling process. Use arrows to match each feature to the correct description. **3 marks**

| Features of coral reef nutrient cycling process | Description |
|---|---|
| Coral polyp | A microscopic algae living in the tissue of a coral polyp. |
| Zooxanthellae | Microscopic organisms found in surface waters that feed on phytoplankton. |
| Zooplankton | A dissolved form of nitrogen excreted by fish. |
| Ammonia | Tiny soft-bodied organism attached to rock. |

4. Explain how nutrients are recycled in coral reefs. **4 marks**

**Top tip:** Draw a quick diagram with arrows showing the flow of nutrients. Label the four key features from Question 3 and annotate four stages to explain the recycling process.

# Coral reefs (Case study 7)

**Coral reef case study:** ......................................................................................................................................

Use the table below to summarise this case study. Use bullet points to list key facts and statistics.

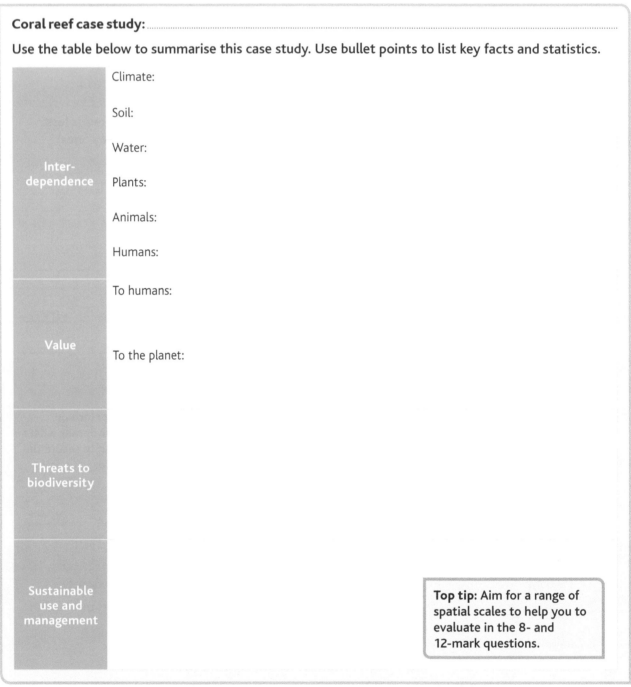

| Inter-dependence | Climate: |
| | Soil: |
| | Water: |
| | Plants: |
| | Animals: |
| | Humans: |
| Value | To humans: |
| | To the planet: |
| Threats to biodiversity | |
| Sustainable use and management | |

**Top tip:** Aim for a range of spatial scales to help you to evaluate in the 8- and 12-mark questions.

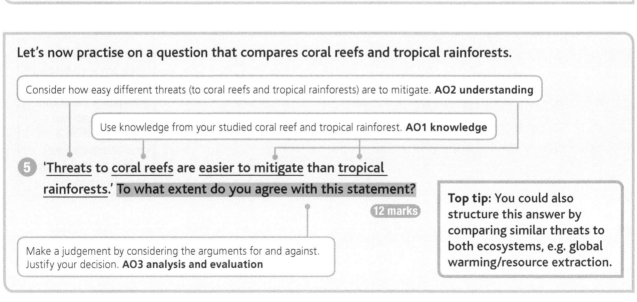

**Let's now practise on a question that compares coral reefs and tropical rainforests.**

Consider how easy different threats (to coral reefs and tropical rainforests) are to mitigate. **AO2 understanding**

Use knowledge from your studied coral reef and tropical rainforest. **AO1 knowledge**

⑤ 'Threats to <u>coral reefs</u> are <u>easier to mitigate</u> than <u>tropical rainforests</u>.' **To what extent do you agree with this statement?**

**12 marks**

Make a judgement by considering the arguments for and against. Justify your decision. **AO3 analysis and evaluation**

**Top tip:** You could also structure this answer by comparing similar threats to both ecosystems, e.g. global warming/resource extraction.

1 **Make your argument that supports the statement.**

Point:
........................................................................
........................................................................

↓

Explain:
........................................................................
........................................................................
........................................................................
........................................................................

↓

Evidence:
........................................................................
........................................................................
........................................................................

↓

Link:
........................................................................
........................................................................
........................................................................
........................................................................
........................................................................

2 **Counter your argument by opposing the statement.**

Point:
........................................................................
........................................................................

↓

Explain:
........................................................................
........................................................................
........................................................................
........................................................................

↓

Evidence:
........................................................................
........................................................................
........................................................................

↓

Link:
........................................................................
........................................................................
........................................................................
........................................................................
........................................................................

**How do I compare the two ecosystems?** Some ideas:

- **Speed** of threat?
- **Longevity** of threat?
- **Size** of threat?
- **Vulnerability** of the ecosystem?

Don't forget to use evaluative connectives!

3 **Conclude** by coming to a judgement. Use the 'opinion line' below to help you.

| I completely agree | I strongly agree | I partially agree | I partially disagree | I strongly disagree | I completely disagree |

**Justification:** ........................................................................
........................................................................

## Want more?

1 Explain the interdependence of abiotic and biotic components in the coral reef you have studied. **8 marks**

2 Assess the statement 'coral reefs are of more value to humans than the planet'. **8 marks**

3 Evaluate the sustainability of the strategies used to mitigate threats in your chosen coral reef. **12 marks**

# Component 2.2: People of the Planet

## Uneven development

You need to know:

- social, economic and environmental definitions of development, including the concept of sustainable development
- how development indicators show the consequences of uneven development across ACs, EDCs and LIDCs and their advantages and disadvantages, particularly for: **GNI per capita**, **Human Development Index** and **internet users**.

---

**1** The table below shows words associated with development. Use arrows to match each key term to the correct definition. **3 marks**

| Development key term | Definition |
|---|---|
| Sustainable development | Investing in the natural world and contributing to its quality and/or quantity. |
| Environmental development | Percentage of the population with access to the internet. |
| GNI/capita | Meeting the needs of the present (socially, economically, and environmentally) while protecting future needs. |
| Human Development Index | Measures life expectancy, education and income per capita, to give a score between 0 and 1. |
| Internet users | Value of all products, taxes and income a country receives including from abroad in a year, divided by the population of the country. |

---

**2** Study the development indicators below. For each indicator use a different coloured highlighter to differentiate whether it is a social, economic or environmental development indicator.

| Key | 🖉 |
|---|---|
| Social | |
| Economic | |
| Environmental | |

Birth rate    $CO_2$ concentrations    HDI

GNI/capita    Infant mortality    Biodiversity

Access to education    Internet users    Literacy rate

> **Top tip:** Some indicators may fit into more than one category, so you may need to highlight the indicator half in one colour, and half in another.

---

**3** State **one** indicator of social development and how it shows how developed a country is. **2 marks**

.........................................................................................................

.........................................................................................................

.........................................................................................................

> **Top tip:** Think about how the indicator shows both low and high levels of development.

---

**4** Study the statements about the three development indicators below.

  a Highlight in one colour advantages, and another colour disadvantages.

  b Then write their number on the Venn diagram according to which development indicator(s) they are referring to.

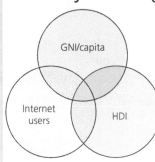

  **1** Data from LIDCs can be unreliable.

  **2** Easy to calculate using official government figures.

  **3** Effectively shows differences between countries and global patterns.

  **4** Does not consider variations within countries and hides inequalities.

  **5** Useful as it relies upon other infrastructure in place so is an interlinked indicator.

  **6** Only an economic indicator, does not show social or environmental development.

  **7** Considers a range of factors and influences on development.

# Consequences and causes of uneven development

There are several causes of uneven development, including the **impact of colonialism on trade and the exploitation of natural resources**. You also need to know **different types of aid and their role in both promoting and hindering development**.

1 Study **Figure 3.5**, which shows the correlation between GNI/capita and life expectancy for 40 countries.

  a  Draw a trend line onto **Figure 3.5**.   `1 mark`

  b  Circle a country on **Figure 3.5** that is an anomaly.   `1 mark`

  c  For the data point in blue, suggest whether it is an AC/ EDC/LIDC and give an example of a country it could be.   `2 marks`

......................................................................................................

......................................................................................................

> **Top tips:**
> - A **line of best fit** = a **straight line** through the data with an equal number of pieces of data above and below the line.
> - **Trend line** = a **curved line** (where the pattern is curved).

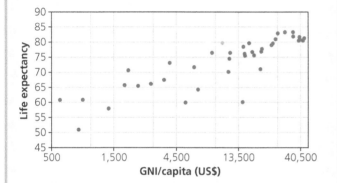

Figure 3.5 Scatter graph showing the correlation between GNI/capita and life expectancy for 40 countries in 2017

## Correlation

On a scatter graph, look for a correlation as outlined in **Figure 3.6** below. If the data points closely follow the line of best fit, it is a 'strong' correlation; if it only loosely follows the line, it is a 'weak' correlation.

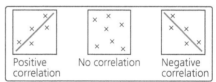

Figure 3.6 Correlation on a scatter graph

2 Explain the role colonialism can have on the development of a country.   `4 marks`

  a  **Read** the answer to the question below. It would get Level 1 (2/4 marks). Why do you think it got this mark?

  b  **Edit** it by correcting any SPaG and annotating any key terms, including connectives to expand the points and examples.

  c  **Improve** it by rewriting it yourself. Include the following terms: **host country, 'mother' country, European powers/British Empire, development**.

> Colonialism is wear a country enters another country and claims power over it. This happened in Africa in the past. They took raw materials for themselves, weather the country liked it or not.
>
> Countrys werent allowed to trade with anyone else and sometimes these trade links were unfair. Some countrys are still impacted by colonialism today.

## Want more?

1 Explain how the exploitation of natural resources can hinder development.   `4 marks`

2 Explain how aid can impact development.   `4 marks`

# Changing economic development (Case study 8)

For one LIDC/EDC you need to know the country's path of economic development through **Rostow's Model**.

Use the table below to summarise this case study. **Use bullet points to list key facts and statistics.**

**LIDC/EDC case study:** ...........................................................................................................................................

| | | |
|---|---|---|
| **Location and environmental context** | Location: | |
| | Landscape: | |
| | Climate: | |
| | Ecosystems: | |
| | Availability and type of natural resources: | |
| **Politics** | Political development: | Relationships with other states: |
| **Economics** | Key imports: | Key exports: |
| | Relative importance of trade: | Role of international investment: |
| **Social factors** | Population structure (+ over time): | Employment structure (+ over time): |
| | Access to education: | Healthcare provision: |
| | Aid project: | |
| **Technology** | Technological developments: | |

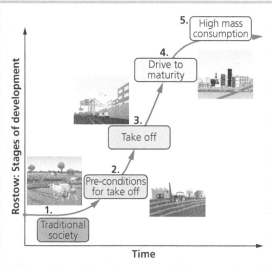

**Figure 3.7 Rostow's model of development**

5. High mass consumption
4. Drive to maturity
3. Take off
2. Pre-conditions for take off
1. Traditional society

Rostow: Stages of development

Time

**1** Study **Figure 3.7**, which shows Rostow's model of development.

Give an account that addresses a range of ideas and arguments. **AO3 analysis and evaluation**

Use knowledge from your studied LIDC/EDC to provide evidence for which stage it is at. **AO1 knowledge**

**Discuss** which <u>stage</u> an <u>LIDC</u> or an <u>EDC you have studied</u> has reached on <u>Rostow's model.</u>　**6 marks**

Consider the stages of Rostow's model of development – which is the best fit stage for your case study?

Aim to argue how your chosen LIDC/EDC fits multiple stages. Then conclude which stage you feel it best fits.
AO2 understanding

Read the example paragraph below, which argues that the DRC is at Stage 2. Pick apart why it is a good paragraph by completing the following steps:

a Underline the connectives it has used for developing sentences further/providing evidence.

b Highlight any evidence provided in the form of facts/statistics.

c Write an 'L' every time the answer makes a clear link back to the question.

The student makes a clear point at the start of the paragraph. The paragraph follows the PEEL structure.

It can be argued the DRC is in Stage 2 (Pre-conditions for economic take off) of Rostow's model for development. In this stage, industry begins to extract natural resources, which can be seen in the DRC where its economy has become heavily dependent on the mining sector. For example, 24% of its GDP and 85% of its export revenue comes from the mining sector. Furthermore, communications develop in Stage 2, which is also evidenced in the DRC with 22 million out of a total population of 82 million using mobile phones. These statistics provide strong evidence the DRC has moved out of Stage 1 into Stage 2.

The answer refers to expected criteria for that stage (underlined) as a reference point by which to judge whether it has reached that stage. **This is important in your answer!**

For your own answer, decide where the country you have studied is on the Rostow model by marking an 'X' on **Figure 3.7** above. Then write an answer to Question 1. Remember to use PEEL!

## Want more?

Pick an option in each question to answer.

**1** Explain how a country's location and environmental context/population and employment structure affects the economic development of an LIDC or and EDC you have studied.　**8 marks**

**2** Analyse the influence of social factors/politics on the economic development of an LIDC or an EDC you have studied.　**8 marks**

# Urban areas

You need to know:

- the definition of a **city**, **megacity** and **world city**
- where megacities are distributed globally and how this has changed over time
- how urban growth rates differ globally with contrasting levels of development.

**1** The table below shows **three** different types of urban areas. Use arrows to match each urban area to the correct description. `2 marks`

| Urban area | Description |
|---|---|
| City | A settlement considered to be an important hub in the global economic system and one that has iconic status and buildings. |
| Megacity | A settlement with a population of over 10 million people. |
| World city | A large human settlement with an extensive system of services and functions. |

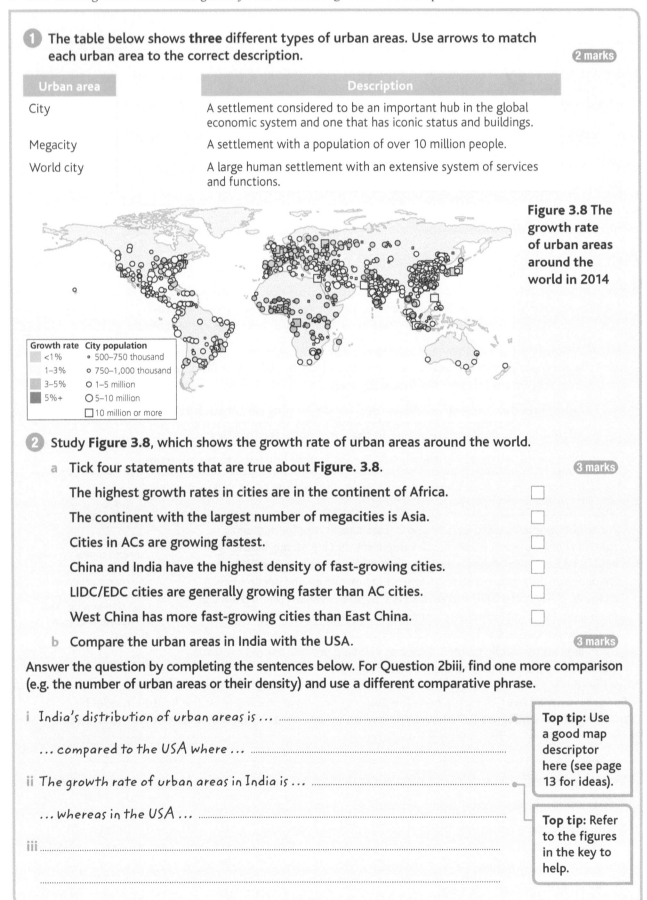

**Figure 3.8 The growth rate of urban areas around the world in 2014**

| Growth rate | City population |
|---|---|
| <1% | • 500–750 thousand |
| 1–3% | ○ 750–1,000 thousand |
| 3–5% | ○ 1–5 million |
| 5%+ | ○ 5–10 million |
| | □ 10 million or more |

**2** Study **Figure 3.8**, which shows the growth rate of urban areas around the world.

a Tick four statements that are true about **Figure. 3.8**. `3 marks`

The highest growth rates in cities are in the continent of Africa. ☐

The continent with the largest number of megacities is Asia. ☐

Cities in ACs are growing fastest. ☐

China and India have the highest density of fast-growing cities. ☐

LIDC/EDC cities are generally growing faster than AC cities. ☐

West China has more fast-growing cities than East China. ☐

b Compare the urban areas in India with the USA. `3 marks`

Answer the question by completing the sentences below. For Question 2biii, find one more comparison (e.g. the number of urban areas or their density) and use a different comparative phrase.

i India's distribution of urban areas is ... ........................................................................

... compared to the USA where ... ........................................................................

ii The growth rate of urban areas in India is ... ........................................................................

... whereas in the USA ... ........................................................................

iii ........................................................................

........................................................................

> **Top tip:** Use a good map descriptor here (see page 13 for ideas).

> **Top tip:** Refer to the figures in the key to help.

# Rapid urbanisation in LIDCs

You need to know the **causes** of this rapid urbanisation (push and pull factors, and natural growth) and the **social**, **economic**, and **environmental consequences**.

**1** **Read the statements below and write the corresponding letter in the correct circle to differentiate between push and pull factors.**

**A** More opportunities for employment than in rural areas.

**B** There are better healthcare systems and schools in urban areas.

**C** Rural areas often have fewer services and poorer infrastructure.

**D** Cities become transport hubs, encouraging new arrivals and drawing in a workforce from the surrounding area.

**E** Wages in rural areas are at poverty levels in many countries.

**F** Stories of people doing better in the city filter back to rural areas.

**G** Prestige comes from a city location.

**H** Crop failure and famine in rural areas.

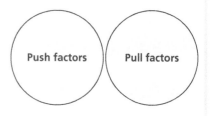

**2** **Explain two consequences of rapid urbanisation in LIDCs.** **4 marks**

**Match up the sentences to answer Question 2. For each consequence use a different coloured highlighter to differentiate whether it is a social, economic or environmental consequence.**

| Key | 🖍 |
|---|---|
| Social | ☐ |
| Economic | ☐ |
| Environmental | ☐ |

**Top tip:** Some indicators may fit into more than one category, so you may need to highlight half the indicator in one colour, and half in another.

'Explain' connectives in bold to explain the significance of the point in the previous column.

Not all consequences are negative – some are positive!

Social/ economic/ environmental consequences of rapid urbanisation in LIDCs are ...

... infrastructure issues such as traffic congestion and adequate housing ...

... the development of slums due to the lack of adequate housing ...

... the World Bank providing extra funding for infrastructure improvements ...

... people are forced to find employment in the informal sector ...

... informal housing is built on marginal land ...

... inadequate energy supplies such as gas and electricity ...

... education is not provided for all the children who arrive ...

... **consequently**, the quality of life is lowered for those settling there.

... **therefore**, they do not pay tax, have no legal rights, and their jobs are less reliable.

... **meaning** cooking is done by wood fires, which lowers local air quality.

... **resulting in** restrictions of further economic growth.

... **as a result**, these properties are vulnerable to flooding and landslides and are located close to industrial activity, which may be bad for residents' health.

... **which** has enabled easier access to some LIDC cities.

... **consequently**, their opportunities to earn a higher income and improve their standard of living later in life are limited.

# Major city in an LIDC or EDC (Case study 9)

**Major city in an LIDC or EDC case study:** ....................................................................................................

Use the table below to summarise this case study. Use bullet points to list key facts and statistics. Be as specific as possible.

| | |
|---|---|
| **Influence of the city on ...** | Region:<br><br>Country:<br><br>Wider world: |
| **Influence of migration (national and international) on the city's ...** | Growth:<br><br>Character: |
| **Influence on ways of life in the city ...** | Culture:<br><br>Ethnicity:<br><br>Housing:<br><br>Leisure:<br><br>Consumption: |
| **Contemporary challenges** | Housing availability:<br><br>Transport provision:<br><br>Waste management: |
| **Sustainable strategies to overcome ...** | |

**Top tip:** You only need to look at the management of **one** challenge.

Offer a reasoned judgement informed by relevant facts. **AO3 analysis and evaluation**

Consider how different ways of life (e.g. culture, lifestyle and consumption, ethnicity and housing) have impacted your city. **AO2 understanding**

Possible '**influence**' on:
- character of the city
- growth of the city
- spatial patterns of the city
- challenges city faces
- population patterns
- employment patterns
- political decisions
- local/regional economy
- local environment.

1 **Assess** the <u>influence of ways of life</u> within a <u>city in an LIDC</u> <u>or an EDC</u>.          `12 marks`

Use knowledge from your studied city in an LIDC/EDC. **AO1 knowledge**

Use the radar diagram below to help answer this question.

a In each box outline a way of life, then explain how it has influenced the city.

b Mark an 'X' on each line on the radar to state the extent the way of life has **influenced** the city.

c Join up your 'Xs' to make a radar diagram.

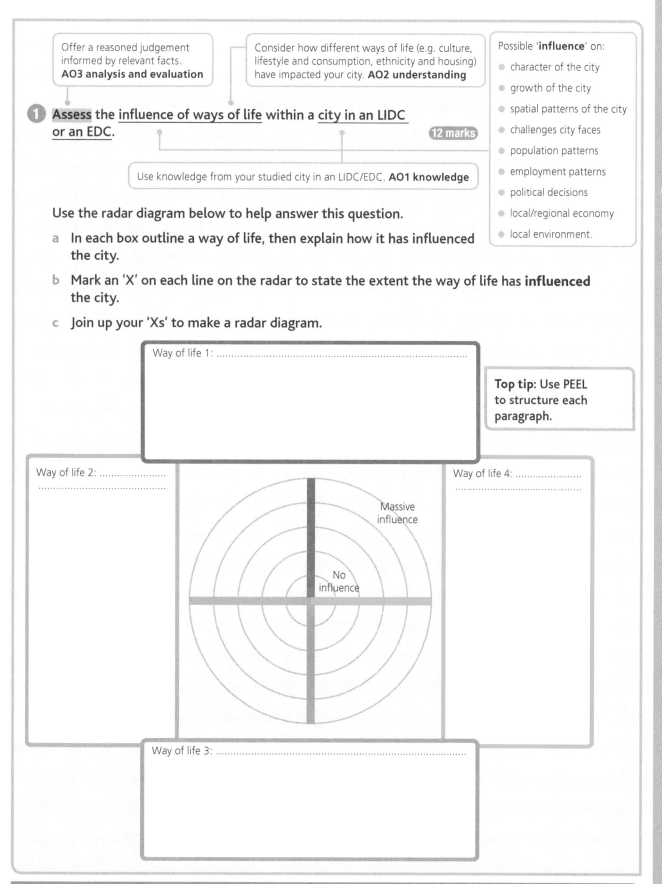

Way of life 1: ...............................................................................

**Top tip:** Use PEEL to structure each paragraph.

Way of life 2: .........................
...............................................

Way of life 4: ........................
..................................................

Massive influence

No influence

Way of life 3: ...........................................................................

## Want more?

1 Explain how the city has influenced areas outside its city boundaries.          `6 marks`

2 Examine the greatest contemporary challenge affecting your chosen city.          `8 marks`

3 Evaluate the sustainability of strategies used in your chosen city to overcome one challenge.          `12 marks`

# Component 2.3: Environmental threats to our Planet

## Climate change over the Quaternary Period

You need to know:

- how the climate has changed, including the ice ages, the medieval warming, the Little Ice Age and modern warming
- evidence for climate change through global temperature data, ice cores, tree rings, paintings and diaries.

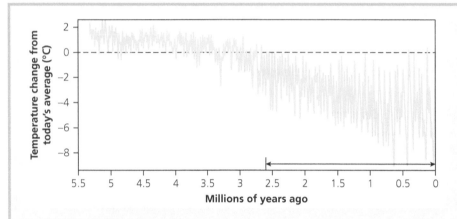

**Figure 3.9 Average global temperatures for the last 5.5 million years**

1. Study **Figure 3.9** which shows the average global temperatures over the last 5.5 million years.

   a. Which of the following statements correctly describes the time period marked by the arrow?

   A  Glacial period                    C  Little Ice Age

   B  Inter-glacial period              D  Quaternary Period

   Write the correct letter in the box.        **1 mark**

   b. Identify **two** features of the average global temperature pattern for the Quaternary period.        **2 marks**

   1 ......................................................................................

   2 ......................................................................................

   > **Top tip:** Use adjectives to help 'identify' the features precisely.

2. The table below shows **three** different time periods during the Quaternary Period. Match each time period to the correct date.        **2 marks**

**Figure 3.10 'The Frost Fair of 1814', on the River Thames in London**

| Time period | Date |
| --- | --- |
| Modern warming | 1300–1870 AD |
| Medieval warming | 1880–present |
| Little Ice Age | 950–1250 AD |

3. Study **Figure 3.10**, which shows the last Frost Fair on the River Thames in London in 1814. Using **Figure 3.10** describe how paintings and diaries can be used as evidence of climate change.        **3 marks**

> **Top tip:** Make sure you refer to **Figure 3.10**!

...................................................................................................................................

...................................................................................................................................

...................................................................................................................................

# Causes of climate change

There are natural and human causes. Natural causes of climate change include variations in energy from the Sun, changes in the Earth's orbit and volcanic activity. More recently, human activity is responsible for the enhanced greenhouse effect which contributes to global warming.

**1** Which of the following statements correctly describes a natural cause of climate change?

    A  Deforestation      C  Burning of fossil fuels

    B  Volcanic activity      D  Decay of waste in landfill sites

    Write the correct letter in the box.        **1 mark**

**2** Explain how humans contribute to global warming.    **4 marks**

**Read the statements below and place them in order (by writing a number next to them) to help answer the question. The first one has been done for you.**

| | | |
|---|---|---|
| **1** Since 1880, human activities have included burning fossil fuels, driving cars, and deforestation. | The enhanced greenhouse gas effect increases global temperatures leading to global warming. | Consequently, the greenhouse gas concentration has increased in the atmosphere. |
| The effect of this is an enhanced effect of the natural greenhouse gas effect in the atmosphere. | These activities have released greenhouse gasses such as carbon dioxide and methane into the atmosphere. | As a result, outgoing long-wave radiation is increasingly trapped in the atmosphere by greenhouse gases. |

**3** **Explain** how **climate change** can be **caused by natural causes.**    **6 marks**

| Provide reasons. | Provide ways climate change has been caused by natural means – NOT human causes! **AO2 understanding** | **Top tip:** 6 marks = 3 causes well explained. |
|---|---|---|

**Complete the table below using the structure provided to help develop your explanations.**

| POINT: Natural causes | EXPLAIN | Connective to explain further | Further explanation (the significance of cause on the climate) |
|---|---|---|---|
| 1. Changes in the Earth's orbit (Milankovitch cycles) | Changes in the Earth's orbit, such as eccentricity every 100,000 years, coincides closely with glacials and inter-glacials in the Quaternary Period ... → | ...consequently... → | → |
| 1. Volcanic activity | → | → | → |
| 2. Variations in energy from the Sun | → | → | → |

# Consequences of climate change

There are a range of consequences of climate change currently being experienced across the planet.

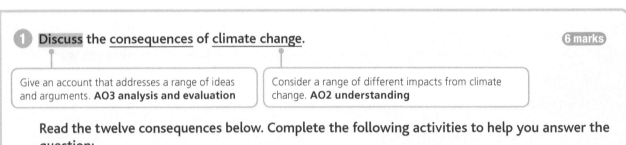

**1** **Discuss** the <u>consequences</u> of <u>climate change</u>. **(6 marks)**

Give an account that addresses a range of ideas and arguments. **AO3 analysis and evaluation**

Consider a range of different impacts from climate change. **AO2 understanding**

Read the twelve consequences below. Complete the following activities to help you answer the question:

a Use three highlighter pens to distinguish between social, economic and environmental consequences.

b Write a '+' for positive and a '−' for negative consequences.

c Write 'S' for short-term, 'M' for medium-term and 'L' for long-term consequences.

d Rank the twelve consequences in order of most to least serious using the numbers 1–12.

e Going further: *Find an example of where each of these consequences are occurring around the world.*

> **Key**
> Social
> Economic
> Environmental

> **Top tip:** The best answers will provide examples.

> **Top tip:** Some indicators may fit into more than one category, so you may need to highlight the indicator half in one colour, and half in another.

**Forest fires**

Increases in temperature mean that large areas of forest catch fire, causing destruction of habitats and sometimes death.

**Wildlife**

Species of birds, animals and insects will have to shift where they live and could become extinct, e.g. the size of sea ice in the Arctic has decreased by 50% since 1979. This has a large effect on species like polar bears.

**Coral reefs**

Coral reefs are an important part of ocean ecosystems, but higher temperatures cause coral polyps to expel algae which results in coral bleaching.

**Sea level rise**

Warming ocean temperatures result in thermal expansion of water and lead to ice caps melting. Sea levels have risen by approximately 19cm between 1901 and 2010.

**Ocean currents**

Climate change could result in increased precipitation and ice melt (e.g. surface area of ice in the Arctic has decreased by 50%). This could alter the strength and position of the ocean currents.

**Farm crops**

In theory, extra $CO_2$ in the atmosphere makes plants grow better so climate change could lead to more food and industrial crops (e.g. biofuel, cotton).

**Permafrost melt**

25% of the Northern hemisphere is covered in permafrost (permanently frozen ground). If it melts, the ground could become soggy. It will also release trapped methane and $CO_2$.

**Extreme weather events: 'global weirding'**

Changes in rainfall patterns lead to flooding in some areas and droughts in others. Increased evaporation may result in more intense tropical storms.

**Health**

Climate change could lead to a greater risk of diseases and ill-health – e.g. from the effects of heavy rainfall or increased numbers of insects like malaria-carrying mosquitos.

**Water supply**

Water is already scarce in many parts of the world. Too little rain and rivers and lakes may dry up. Shrinking glaciers will reduce water for communities at the base of mountains such as the Himalayas.

**Population migration**

Climate change could mean that certain areas become uninhabitable, forcing large numbers of people to migrate.

**Longer growing seasons**

More northerly regions of the world may experience longer growing seasons for crops.

**Now use your work in the table to answer the question on a separate piece of paper.**

# Global circulation of the atmosphere

You need to know:

- how it is controlled by the movement of air between the poles and the Equator
- how it leads to extreme weather conditions (wind, temperature, precipitation) around the world
- the distribution of the main climatic regions of the world.

**1** Study **Figure 3.11**, which shows the six climate zones.

Identify the climatic zone at **X**.

A Polar          C Arid

B Temperate          D Tropical

Write the correct letter in the box.          **1 mark**

To help answer the question write the correct climate zone next to the corresponding colour in the key on **Figure 3.11**. The six climate zones are: **Mediterranean**, **Temperate**, **Polar**, **Mountains**, **Arid**, **Tropical**.

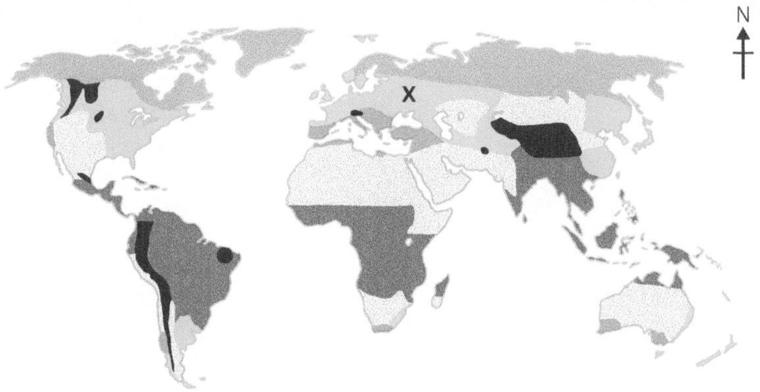

N

**Figure 3.11 Map showing six climate zones**

Key

**Top tip:** The Hadley Cell can be explained using an annotated diagram. For three marks aim to describe three roles.

**2** Explain the role of the Hadley Cell in moving air around the planet.          **3 marks**

Use a diagram to help you answer Question 2 on a separate piece of paper.

**3** Explain how parts of the <u>global circulation of the atmosphere</u> lead to extreme weather conditions such as precipitation.          **3 marks**

> Just focus on the role of the global circulation of the atmosphere!

Complete the table to answer Question 3.

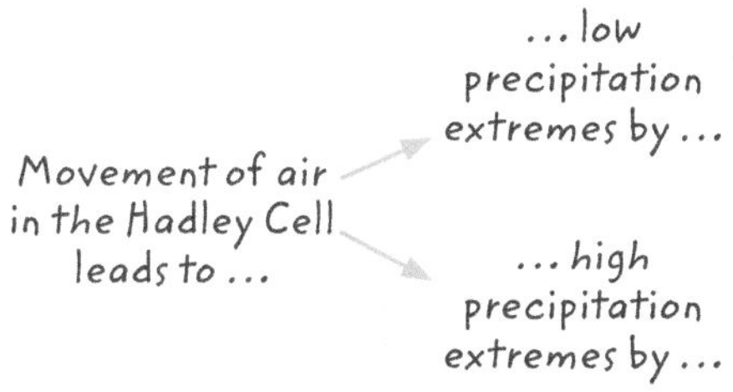

Movement of air in the Hadley Cell leads to ...

→ ... low precipitation extremes by ...

→ ... high precipitation extremes by ...

## Going further

**1** Explain how parts of the global circulation of the atmosphere lead to extreme weather conditions such as temperature/wind.          **3 marks**

# Extreme weather – tropical storms

For tropical storms and drought, you need to know the **causes**, their **distribution** and **frequency**, and **whether they have changed over time**.

**1** Study **Figure 3.12** which shows the cross-section of a hurricane in the Northern Hemisphere. Use **Figure 3.12** to explain how tropical storms can cause extreme weather. **4 marks**

........................................................................................................

........................................................................................................

........................................................................................................

........................................................................................................

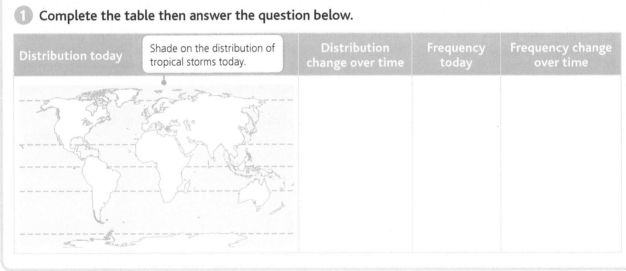

Figure 3.12 Cross-section of a hurricane in the Northern hemisphere

**Top tip:** Pick out features in **Figure 3.12** to help explain how tropical storms CAUSE extreme weather, e.g. *warm rising air cools, condenses and forms tall cumulonimbus clouds resulting in ...*

**1** Complete the table then answer the question below.

| Distribution today Shade on the distribution of tropical storms today. | Distribution change over time | Frequency today | Frequency change over time |
|---|---|---|---|
| | | | |

**2** Outline how the frequency of tropical storms has changed over time. **3 marks**

........................................................................................................

........................................................................................................

........................................................................................................

**Top tip:** The answer is complicated. Acknowledge that it differs around the world. You can use phrases such as: '*The evidence suggests ...*'

# Extreme weather – drought

**1** Define a drought. ............................................................................................ **1 mark**

.............................................................................................................................

**2** Explain the causes of a drought. **4 marks**

Use the table below to answer the question. The first cause has been done for you.
Complete two more causes – make sure they are different from on another.

| POINT: (causes) | EXPLAIN: | Connective to explain further | Further explanation (the significance of the cause in creating a drought) |
|---|---|---|---|
| 1. **Physical cause:** Increase in global temperatures ... | ... means more water is needed to grow crops and more water is lost through evaporation ... | ... consequently ... | ... soils and surface water supplies deplete and dry out. |
| 2. **Physical cause:** | → | → | → |
| 3. **Human cause:** | → | → | → |

**1** Complete the table on drought, then answer the questions below.

| Distribution today | Shade on the distribution of droughts today. | Distribution change over time | Frequency today | Frequency change over time |
|---|---|---|---|---|
| | | | | |

**3** Outline how the distribution of droughts has changed over time. **3 marks**    **Top tip:** Refer to specific locations.

.............................................................................................................................

.............................................................................................................................

.............................................................................................................................

.............................................................................................................................

# Drought (Case study 10)

## Drought case study:

Use the table below to summarise this case study. Use bullet points to list key facts and statistics.

| | |
|---|---|
| **How El Niño/ La Niña develops and can lead to drought** | |
| **Effects of drought** | On people:<br><br>On the environment: |
| **Ways people in the area have adapted to drought** | |

**Top tip:** Aim to have examples of adaptations on a range of spatial scales (national/regional/local) and temporal scales (short/ long term).

---

Consider impacts on both people and environment and how effective adaptations to these impacts have been. One paragraph on people, one on environment. **AO2 understanding**

**4** '<u>Impacts on people</u> have been <u>more effectively adapted</u> to than <u>impacts on the environment.</u>' <u>How far do you agree</u> with this statement for <u>one drought event you have studied</u>? **12 marks**

Make a judgement by considering the arguments for and against. Justify your decision. **AO3 analysis and evaluation**

Use knowledge from your studied drought from an El Niño/La Niña event. **AO1 knowledge**

You are now going to answer this question on page 73.

Use PEEL to structure each paragraph. When you get to 'L' you need to think critically as to how effective (see box) the strategies in your paragraph have been.

**How do I consider effectiveness?**
Consider:
- **speed** of adaptation
- **scale** of adaptation
- **longevity** of adaptation
- **vulnerability** of people/environment.

**1 Make your argument** that supports the statement (impacts on people more effectively adapted to).

..............................................................................................................................................................................
..............................................................................................................................................................................
..............................................................................................................................................................................
..............................................................................................................................................................................
..............................................................................................................................................................................

**2 Counter your argument** by opposing the statement (impacts on environments more effectively adapted to).

..............................................................................................................................................................................
..............................................................................................................................................................................
..............................................................................................................................................................................
..............................................................................................................................................................................
..............................................................................................................................................................................

**3 Conclude** by using the 'opinion line' below.

| I completely agree | I strongly agree | I partially agree | I partially disagree | I strongly disagree | I completely disagree |

**Justification:**

..............................................................................................................................................................................
..............................................................................................................................................................................
..............................................................................................................................................................................

## Want more?

**1** Explain how an El Niño/La Niña event can lead to drought. **6 marks**

**2** 'The consequences of climate change are more positive than negative.' Assess this statement. **8 marks**

# Chapter 4 Tackling Paper 3: Geographical Skills and Fieldwork Assessment

## Section A: Geographical Skills

Most questions you will come across in Section A are worth 1, 2, 3 or 4 marks. These will examine your cartographic, graphical, numerical and statistical skills. Aim to spend about 45 minutes on this section.

### OS maps

For OS maps you need to be able to:

- use and understand coordinates, scale and distance
- extract, interpret, analyse and evaluate information
- understand gradient, contours and spot heights, and be able to use them to draw cross-sections
- use the map to draw sketch maps.

> **Top tip:** Pages 13–14 of the OCR A Specification outlines all the geographical skills you could be examined on. Use the list on these pages as a checklist to ensure you have revised each one.

---

1. Study **Figure 4.1** on page 75, an OS map extract of Kirkby Moor, south of the Lake District.

   a **Give the 4-figure grid reference for the six wind turbines on Kirkby Moor.** ........................................................................................ (1 mark)

   b **What is the highest point on Kirkby Moor?** •

   | | | Write the correct letter in the box. ☐ (1 mark) |
   |---|---|---|
   | A 315m | C 333m | |
   | B 332m | D 357m | |

   > **Top tip:** You are looking for a spot height in black.

   c **In which general direction would you be travelling if you were approaching Foxfield from the Kirkby-in-Furness by train?** • (1 mark)

   ........................................................................................

   > **Top tip:** Be careful to get the direction the right way round!

   d **What is the distance between Foxfield and Kirkby-in-Furness railway stations?** • (1 mark)

   ........................................................................................

   > **Top tip:** See page 11 for how to do this. Don't forget the correct unit at the end!

   e **Suggest two possible sources of income from Kirkby Moor for the residents of Kirkby-in-Furness.** (2 marks)

   ........................................................................................

   ........................................................................................

   ........................................................................................

   ........................................................................................

   ........................................................................................

   ........................................................................................

   ........................................................................................

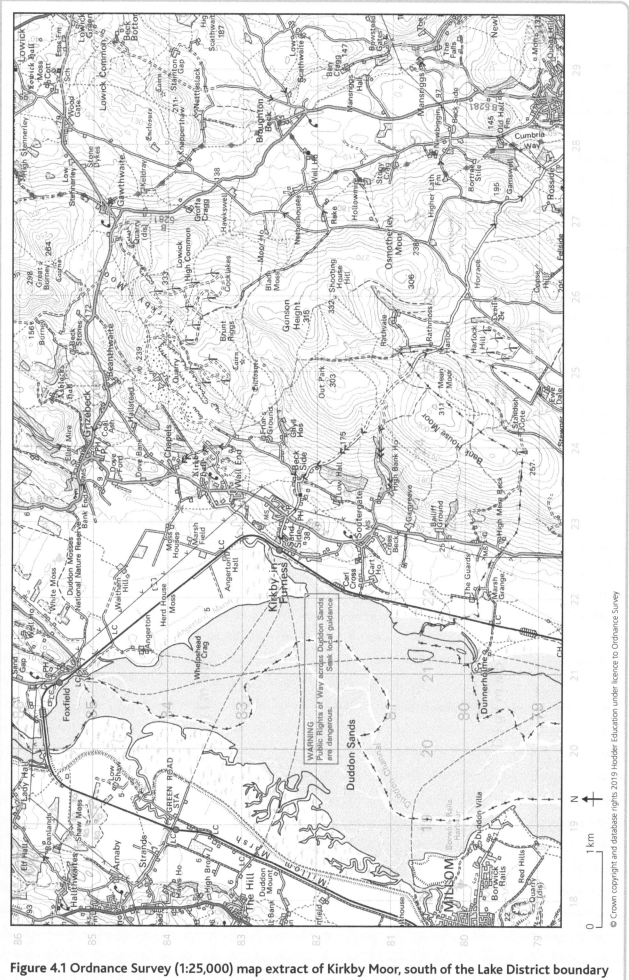

**Figure 4.1** Ordnance Survey (1:25,000) map extract of Kirkby Moor, south of the Lake District boundary

# Other maps

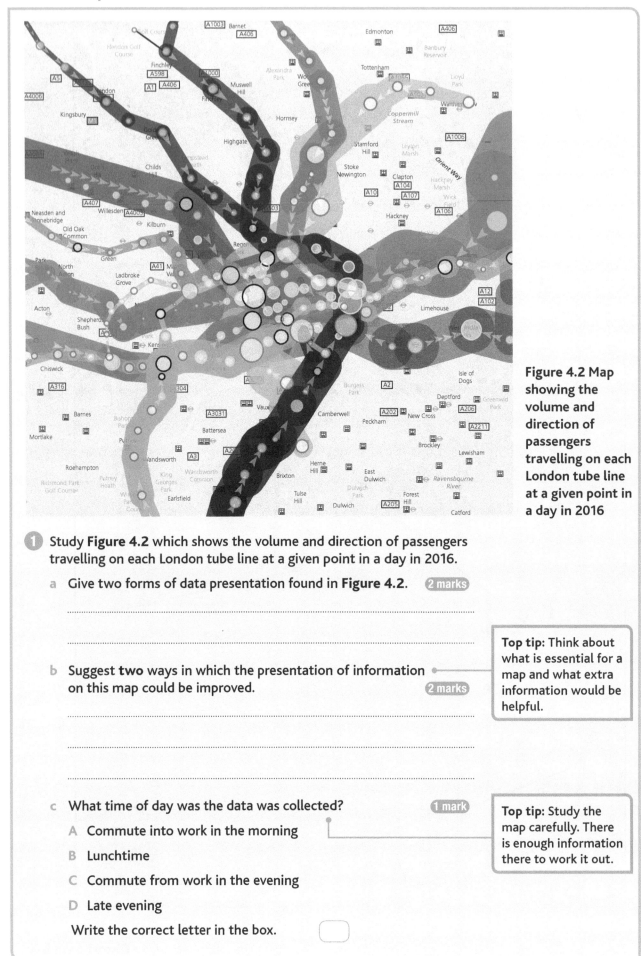

**Figure 4.2** Map showing the volume and direction of passengers travelling on each London tube line at a given point in a day in 2016

1 Study **Figure 4.2** which shows the volume and direction of passengers travelling on each London tube line at a given point in a day in 2016.

a Give two forms of data presentation found in **Figure 4.2**. **2 marks**

.................................................................................................................

.................................................................................................................

b Suggest **two** ways in which the presentation of information on this map could be improved. **2 marks**

> **Top tip:** Think about what is essential for a map and what extra information would be helpful.

.................................................................................................................

.................................................................................................................

.................................................................................................................

c What time of day was the data was collected? **1 mark**

> **Top tip:** Study the map carefully. There is enough information there to work it out.

A Commute into work in the morning

B Lunchtime

C Commute from work in the evening

D Late evening

Write the correct letter in the box.  ☐

d Suggest a reason for you answer to 1c. **2 marks**

...............................................................................................................

...............................................................................................................

...............................................................................................................

e Which urban trend does **Figure 4.2** suggest the London tube lines have made possible? **1 mark**

A Urbanisation

B Counter-urbanisation

C Suburbanisation

D Re-urbanisation

Write the correct letter in the box.

f Describe the pattern of movement of passengers on the London tube lines. Refer to **Figure 4.2** in your answer. **4 marks**

> **Top tip:** There's no numerical data, but you can refer to the colours, direction and thickness of the lines.

...............................................................................................................

...............................................................................................................

...............................................................................................................

...............................................................................................................

...............................................................................................................

...............................................................................................................

...............................................................................................................

...............................................................................................................

...............................................................................................................

g Using **Figure 4.2** and your <u>own knowledge</u>, explain reasons for the geographical patterns you can see. **3 marks**

> **Top tip:** This links to something you can apply from the UK or World paper. Hint: *People of the UK topic!*

...............................................................................................................

...............................................................................................................

...............................................................................................................

...............................................................................................................

...............................................................................................................

...............................................................................................................

...............................................................................................................

...............................................................................................................

# Graphs

**1** Precipitation data for Betws-y-Coed, Wales, is shown in the table below.

|  | Jan | Feb | Mar | Apr | May | Jun | Jul | Aug | Sep | Oct | Nov | Dec |
|---|---|---|---|---|---|---|---|---|---|---|---|---|
| Precipitation (mm) | 290 | 224 | 243 | 152 | 132 | 133 | 143 | 181 | 209 | 300 | 295 | 310 |

a Complete the climate graph for Betws-y-Coed below using information from the table. **2 marks**

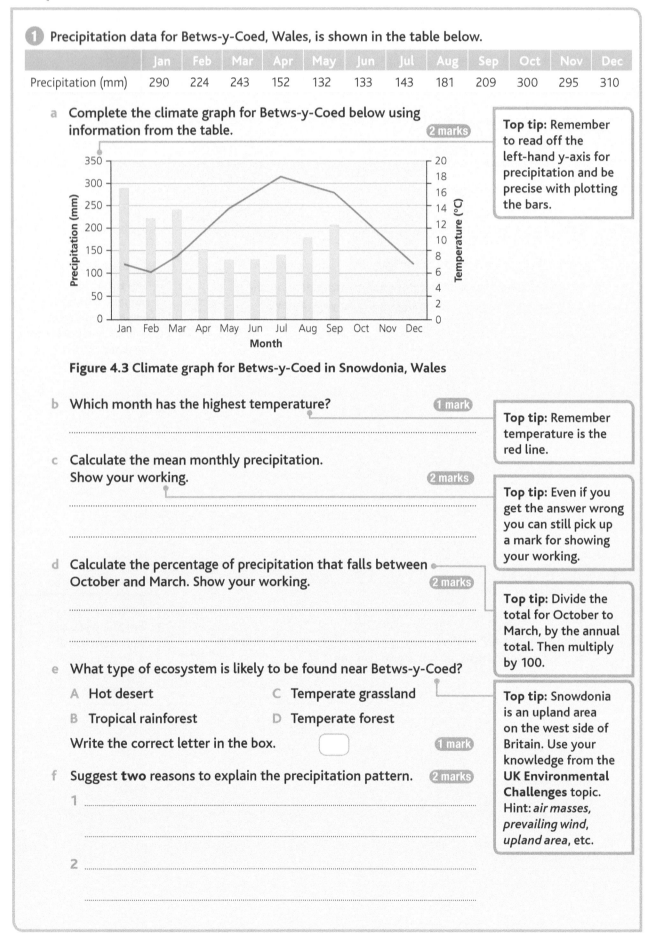

**Top tip:** Remember to read off the left-hand y-axis for precipitation and be precise with plotting the bars.

Figure 4.3 Climate graph for Betws-y-Coed in Snowdonia, Wales

b Which month has the highest temperature? **1 mark**

......................................................................................

**Top tip:** Remember temperature is the red line.

c Calculate the mean monthly precipitation. Show your working. **2 marks**

......................................................................................

......................................................................................

**Top tip:** Even if you get the answer wrong you can still pick up a mark for showing your working.

d Calculate the percentage of precipitation that falls between October and March. Show your working. **2 marks**

......................................................................................

......................................................................................

**Top tip:** Divide the total for October to March, by the annual total. Then multiply by 100.

e What type of ecosystem is likely to be found near Betws-y-Coed?

A Hot desert     C Temperate grassland

B Tropical rainforest     D Temperate forest

Write the correct letter in the box.　　◯　**1 mark**

**Top tip:** Snowdonia is an upland area on the west side of Britain. Use your knowledge from the **UK Environmental Challenges** topic. Hint: *air masses, prevailing wind, upland area,* etc.

f Suggest **two** reasons to explain the precipitation pattern. **2 marks**

1 ......................................................................................

......................................................................................

2 ......................................................................................

......................................................................................

## Tables

**1** Study the table below, which shows the top ten countries for the number of people affected by drought per year between 1999 and 2008.

| India | China | Iran | Kenya | Ethiopia | Thailand | South Africa | Malawi | Zimbabwe | Niger |
|---|---|---|---|---|---|---|---|---|---|
| 35,000,000 | 19,666,000 | 3,700,000 | 3,000,000 | 2,650,000 | 2,100,000 | 1,500,000 | 845,000 | 810,000 | 658,000 |

a State one way this data could be presented on a graph. **1 mark**

..................................................................................................

> **Top tip:** See page 7 for ideas. Hint: *There is only one variable, and you also want to display the country name.*

b State one advantage of presenting this data using that technique. **2 marks**

..................................................................................................

c State one way this data could be presented on a map. **1 mark**

..................................................................................................

> **Top tip:** Be precise with your point, e.g. *you can compare the number of people affected between countries easily.*

d State one advantage and one disadvantage of presenting this data using that technique. **2 marks**

..................................................................................................

..................................................................................................

..................................................................................................

> **Top tip:** See page 10 for ideas.

e Suggest two issues with this data. **2 marks**

1 ..............................................................................................

2 ..............................................................................................

> **Top tip:** Hint: *Up-to-date? Bias? Comparable?*

f In 2008, Kenya's population was estimated at 37,000,000. Between 1999 and 2008, 3,000,000 people were affected by drought per year in Kenya. What is the ratio 37,000,000:3,000,000 in its simplest form? **1 mark**

..................................................................................................

..................................................................................................

> **Top tip:** Don't panic about the large numbers – you can take them out as there are the same number on each side of the ratio.

g Describe the pattern of the number of people affected by drought per year between 1999 and 2008. Use data from the table in your answer. **4 marks**

..................................................................................................

..................................................................................................

..................................................................................................

..................................................................................................

..................................................................................................

..................................................................................................

> **Top tip:** Can you spot any spatial patterns? Think about where are these countries located in the world.

## Photos

**Figure 4.4 Photo taken in Norway**

Similar coloured vegetation suggests low biodiversity.

**Top tip:** Draw your lines accurately to the exact spot on the photo. See page 14 for a reminder about the difference between a label and annotation.

1 Study **Figure 4.4**, which is a photo taken in Norway.

   a Identify **three** pieces of evidence that suggest this is a tundra ecosystem by annotating onto the photo. One annotation has already been done as an example. **3 marks**

   b Using **Figure 4.4** and your own knowledge, explain why the population density is sparse in the Arctic tundra. **3 marks**

........................................................................................................

........................................................................................................

........................................................................................................

........................................................................................................

2 Study **Figure 4.5**, which is a photo of Kirkby Moor from **Figure 4.1** (see page 75).

**Figure 4.5 Kirkby Moor**

   a Which energy source is being used in **Figure 4.5**?

   A Solar energy    C HEP    Write the correct letter in the box.  **1 mark**

   B Wind energy    D Coal

   b Using **Figure 4.1** (page 75) and **Figure 4.5**, identify in which direction the camera was pointing when the photograph was taken.

   A East    C South East    Write the correct letter in the box.  **1 mark**

   B North East    D South West

# Synoptic questions

Within Section A there will be typically **two** synoptic questions (probably **one 6-mark question** and **one 8-mark question**). You will need to draw together knowledge, understanding and skills from different parts of your whole GCSE course and apply it in a different context, e.g. between the UK and World Paper. One of your questions will likely also include a **source** to use.

> ## Top tips for synoptic questions
>
> 1. Use facts to show off your knowledge from different parts of the course to support your points.
>
> 2. Any analysis/evaluation you can bring in to compare the two parts of the question will earn you credit. Here are some comparison ideas to help you:
>    a. compare spatial scales (local/regional/national/international, etc.)
>    b. compare temporal scales (short/medium/long term, etc.)
>    c. compare different levels of economic development
>    d. compare demographics (young/elderly/male/female, etc.)
>    e. compare impacts (social/economic/environmental/political)
>    f. compare the severity of the impacts
>    g. compare the speed or rate of change of impacts
>    h. compare the vulnerability of a place and its ability to manage/adapt to impacts.
>
> 3. Practise making synoptic links between the overlapping themes and models outlined below.

## Overlapping themes

You can do some preparation by identifying overlapping themes. Make comparisons between the UK and World Paper by planning answers in the spaces provided to the questions on the overlapping themes below.

> 1. **Migration** (national and international for UK city and LIDC/EDC city case studies)
>
>    e.g. 'The impact of migration is greater in the UK than in LIDCs/EDCs.' To what extent do you agree with this statement? **8 marks**

> 2. **Urban trends** (suburbanisation, counter-urbanisation and re-urbanisation in the UK and rapid urbanisation in LIDCs)
>
>    e.g. 'UK cities have benefited more from movement of people into them than LIDC cities.' How far do you agree with this statement? **8 marks**

**3** **Ways of life in a city** (such as culture, ethnicity, housing, leisure and consumption in UK and LIDC/EDC city case studies)

e.g. 'Ways of life differ between UK and LIDC/EDC cities.' Discuss. **6 marks**

**4** **Contemporary urban challenges** (housing availability, transport provision and waste management in UK and LIDC/EDC city case studies)

e.g. 'Cities in the UK face challenges which are less serious than cities in LIDCs or EDCs.' To what extent do you agree with this statement? **8 marks**

**5** **Uneven development** (causes in UK and across LIDCs/EDCs)

e.g. 'Causes of uneven development are the same.' Discuss with reference to both the UK and an LIDC/EDC you have studied. **6 marks**

**6** **Sustainable strategies** (to overcome urban challenges, mitigate threats to biodiversity in coral reefs and tropical rainforests, use of energy in UK)

e.g. 'Cities in the UK have more sustainable strategies to overcome challenges than cities in LIDCS/EDCs.' To what extent do you agree with this statement? **8 marks**

**7** **Extreme weather events** (causes, impacts and management in UK, e.g. flooding, and the world, e.g. tropical storms and droughts)

e.g. 'Examine the view that other countries manage extreme weather events better than the UK.' **8 marks**

## Applying models

Questions may also examine your ability to apply a model you have learnt on one topic in a different and unfamiliar context. For example:

- Rostow's Model for the UK
- Demographic Transition Model for the LIDC country you studied/a country other than the UK
- Population pyramid for a country other than the UK

They may also examine your ability to apply your knowledge to unfamiliar contexts within the UK or World Paper, e.g. urban issues in rural contexts.

### Synoptic question

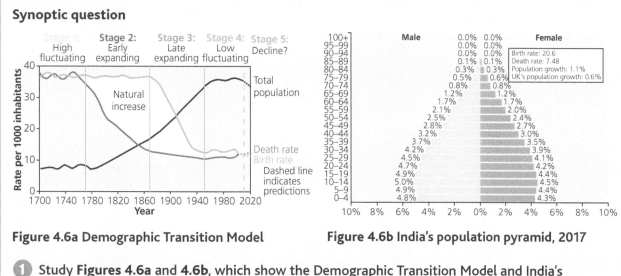

**Figure 4.6a Demographic Transition Model**

**Figure 4.6b India's population pyramid, 2017**

**1** Study **Figures 4.6a** and **4.6b**, which show the Demographic Transition Model and India's population pyramid.

Using Figures 4.6a and 4.6b **discuss** which stage on the DTM you think India is on. **(6 marks)**

| You must make explicit reference to these, e.g. quoting the birth rate. | Give an account that addresses a range of ideas and arguments on the stage of India on the DTM. **AO3 analysis and evaluation** | Need to understand how, as a country develops, it moves through the stages of the Demographic Transition Model and apply that to India, using the information from the population pyramid on India. **AO2 understanding** |

Complete the table below using evidence from **Figures 4.6a** and **4.6b** and your own knowledge of the UK as a comparison. The first row has been done for you.

| Evidence from Figures 4.6a and b | Own knowledge | Link to stage on DTM |
| --- | --- | --- |
| India's population growth rate i 1.1% UK's population growth rate i 0.6% | UK is at Stage 4 In recent years India's population has grown rapidly, catching up with China's. | UK is at Stage 4, therefore because India's population growth rate is higher than the UK's, it would suggest India is at Stage 3 on Figure 4.6a where the total population is growing at a faster rate. |

**Top tip:** Don't forget to use explain connectives.

**Top tip:** Link back to which stage on Figure 4.6a you think India is on.

2 'UK cities have benefited more from movement of people into them than LIDC cities.' How far do you agree with this statement?

**8 marks**

**1 Make your argument** that supports the statement (UK cities benefited more from re-urbanisation).

...............................................................................

...............................................................................

...............................................................................

...............................................................................

...............................................................................

...............................................................................

**Top tip:** Positives of re-urbanisation in UK cities include: *new homes, new jobs, increased investment into areas, etc.*

**2 Counter your argument** by opposing the statement (LIDC cities benefited more from rapid urbanisation).

...............................................................................

...............................................................................

...............................................................................

...............................................................................

...............................................................................

...............................................................................

**Top tip:** Positives of rapid urbanisation in LIDC cities include: *infrastructure improvements, economic growth, etc.*

**3 Conclude** by using the 'opinion line' below. Make sure you justify your decision.

| I completely agree | I strongly agree | I partially agree | I partially disagree | I strongly disagree | I completely disagree |

**Justification:**

...............................................................................

...............................................................................

...............................................................................

# Section B: Fieldwork Assessment

There are two sets of questions about fieldwork in Section B.

- First there are **generic fieldwork questions**, which will contain fieldwork material from an unfamiliar context.
- Second, there are **questions on one of your two individual fieldwork enquiries**. You will have to answer questions on what you did and why.

## Unfamiliar fieldwork

**1** Some geography students have been conducting **physical geography** fieldwork on a beach in the UK. They were investigating pebble size. Their results are shown in **Figure 4.7** below.

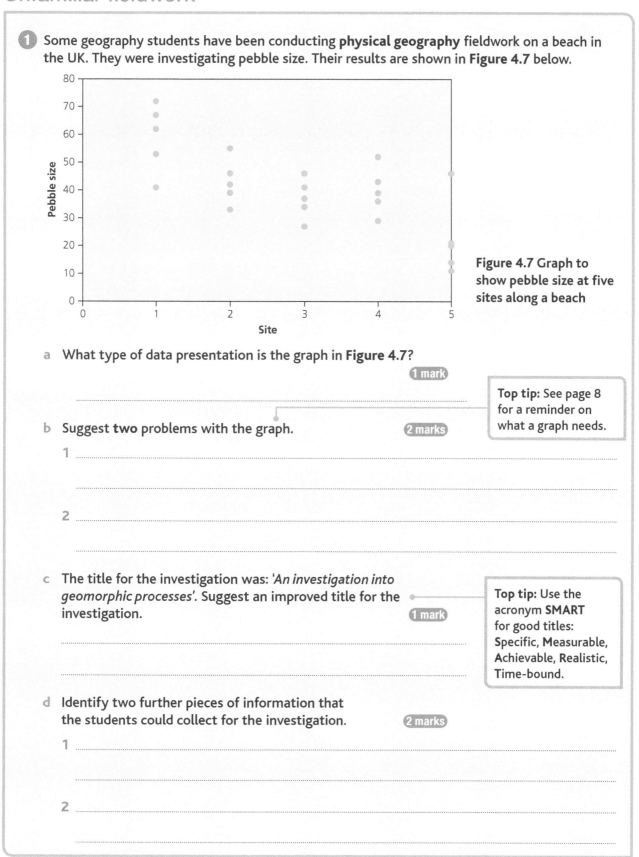

**Figure 4.7** Graph to show pebble size at five sites along a beach

a What type of data presentation is the graph in **Figure 4.7**?

**1 mark**

........................................................................................................

> **Top tip:** See page 8 for a reminder on what a graph needs.

b Suggest **two** problems with the graph. **2 marks**

1 ........................................................................................................

........................................................................................................

2 ........................................................................................................

........................................................................................................

c The title for the investigation was: '*An investigation into geomorphic processes*'. Suggest an improved title for the investigation. **1 mark**

........................................................................................................

........................................................................................................

> **Top tip:** Use the acronym **SMART** for good titles: Specific, Measurable, Achievable, Realistic, Time-bound.

d Identify two further pieces of information that the students could collect for the investigation. **2 marks**

1 ........................................................................................................

........................................................................................................

2 ........................................................................................................

........................................................................................................

e A student has concluded that '*coastal processes are causing a movement of sediment across the beach from Site 1 to Site 5 on the beach*'. <u>Using Figure 4.7</u>, <mark>assess</mark> the <u>evidence for this conclusion.</u>

**6 marks**

Use Figure 4.6 to 'assess' whether the evidence matches up with the conclusion. **AO2 understanding**

Make a reasoned judgement informed by relevant facts. **AO3 analysis and evaluation**

**Use the table below to plan an answer. One row has been done for you.**

**Top tip:** Start with the overall pattern.

| Evidence from Figure 4.7 | How does it support/not support the conclusion? |
| --- | --- |
| The average pebble size is larger at Site 4, than Site 3. | The pebble size does not uniformly decrease between Site 1 and 5. This could be due to a groyne stopping longshore drift, or it could mean different rock types are present along the beach. Consequently, the pattern in Figure 4.7 may not be explained by movement, instead by the rate of erosion for the different rock types. |

**Top tip:** Refer to any anomalies.

f Suggest two possible reasons for the conclusion you identified in 1e.

**4 marks**

**2** Some geography students have been conducting **human geography** fieldwork in a town in the UK. They were investigating the use of different types of solar panels. Their results are shown below.

| | Green Street | Brooklands Road | Vine Street | Victoria Road | Beaumont Avenue | Queen Street | Belville Gardens |
|---|---|---|---|---|---|---|---|
| Thin film | 1 | 8 | 6 | 0 | 3 | 6 | 3 |
| Monocrystalline | 3 | 2 | 0 | 1 | 12 | 1 | 0 |
| Polycrystalline | 1 | 2 | 2 | 0 | 2 | 0 | 1 |

**Figure 4.8** Types of solar panels used in a town in the UK

a Suggest **one** problem with the recording of this data.

(1 mark)

....................................................................................................................................

....................................................................................................................................

b Calculate the mean value for the number of solar panels on each street. You should show your working.

(2 marks)

Mean value is: ...................................

c Describe, in detail, **one** way the data in the table could be presented in a fieldwork investigation. Give reasons for your choice.

(6 marks)

You need to do three things to answer this question. Use the table below to plan your answer.

| Identify the presentation method | Describe how you would present the data in detail | Reasons for this method |
|---|---|---|
| | | 1. |
| | | 2. |
| | | 3. |

Whenever a location is given, consider a method that shows the location.

Be precise and detailed. Imagine you were describing the method to someone who had never seen it before! Hint: *How will you plot the data? What will go on each axis and in the key? Use of colour?*

Think about the strengths of the method, and the patterns it can show.

d   **Suggest a conclusion** that the students might reach for the enquiry question '*How do patterns of solar panel usage vary?*' **Analyse** the evidence from the table to explain how you have reached that conclusion.                                    **6 marks**

> Put forward a conclusion, then discuss the evidence that supports that conclusion.
> **AO3 analysis and evaluation**

> Use the table as evidence for supporting the conclusion you have made.

To answer this question, use the following flow diagram:

State your conclusion → Analyse the evidence in the table to explain how you reached it → Consequently, state your overall **confidence** in the conclusion

> A good conclusion always links to the confidence it has in its conclusion.

..............................................................................................................................

..............................................................................................................................

..............................................................................................................................

..............................................................................................................................

..............................................................................................................................

..............................................................................................................................

..............................................................................................................................

..............................................................................................................................

..............................................................................................................................

..............................................................................................................................

..............................................................................................................................

..............................................................................................................................

e   Suggest two possible reasons for the conclusion you identified in 1d.                **4 marks**

..............................................................................................................................

..............................................................................................................................

..............................................................................................................................

..............................................................................................................................

..............................................................................................................................

..............................................................................................................................

..............................................................................................................................

..............................................................................................................................

..............................................................................................................................

..............................................................................................................................

# Your own fieldwork (human geography)

You will be examined on either your human or physical geography fieldwork. The process of enquiry involves **six stages** – you need to be familiar with all six as you will be **examined on at least two** of them!

**Stage 1:** Understanding of the kinds of question capable of being investigated through fieldwork and an understanding of the geographical enquiry processes appropriate to investigate these.

**Stage 2:** Understanding of the range of techniques and methods used in fieldwork, including observation and different kinds of measurement.

**Stage 3:** Processing and presenting fieldwork data in various ways including maps, graphs and diagrams.

**Stage 4:** Analysing and explaining data collected in the field using knowledge of relevant geographical case studies and theories.

**Stage 5:** Drawing evidenced conclusions and summaries from fieldwork data.

**Stage 6:** Reflecting critically on fieldwork data, methods used, conclusions drawn, and knowledge gained.

Over the next five pages each question is colour-coded to show which stage it is examining.

Several questions require you to evaluate your enquiry process (Stage 6). This involves looking back at the previous five stages and reflecting critically on what went well and how your process of enquiry could have been improved. Complete the five boxes below to help you evaluate your own human fieldwork.

**Stage 1:** In what ways was your fieldwork question SMART/not SMART?

> **Top tip:** Think about both your **sampling strategy** and **methods** to collect the data itself.

**Stage 2:** In what ways were your data collection methods accurate/not accurate?
In what ways were your data collection methods reliable/not reliable?

> **Top tip:** Could you have improved your data presentation method or presented the data differently?

**Stage 3:** In what ways were your data presentation methods clear/not clear?

> **Top tip:** Did your results support your hypotheses?

**Stage 4:** In what ways were your data effective/not effective in helping reach a conclusion?

> **Top tip:** Think about your methodology, data presentation and results.

**Stage 5:** What were the reasons you were confident/not confident in your conclusions? How was your understanding of the topic developed as a result of the investigation?

> **Top tip:** Link back to your original question and any supporting geographical theory.

---

**1** You will have taken part in fieldwork in a **human geography** environment as part of your studies. Examples might include a city or village study.

State your fieldwork question for investigation: ......................................

..................................................................................................

> **Top tip:** You will be asked for your question so learn it!

Location of study area:

..................................................................................................

a  Suggest **one** reason why this was a suitable location to study.   **1 mark**

..................................................................................................

b  Suggest **one** strength of your fieldwork question and **one** way it could have been improved.   **2 marks**

..................................................................................................

..................................................................................................

..................................................................................................

..................................................................................................

> **Top tip:** 'Location' is related to logistics of the fieldwork e.g. *safety, risk assessment, accessibility*, etc.

> **Top tip:** Use **SMART** (see page 85) to help you think about strengths and improvements of your question.

c  Describe and justify **one** technique you used to present your data.   **2 marks**

**Here is an exemplar, with the description and justification separated:**

**Description:** I geo-located my traffic count data by drawing proportional circles onto a map of the town using ArcGIS software.

**Justification:** Consequently, the size of the circles and their location enabled me to easily identify any spatial patterns for traffic.

**Now answer this question yourself.**

..................................................................................................

..................................................................................................

..................................................................................................

..................................................................................................

'*' means there are 3 SPaG marks

Give an account of your fieldwork. Don't spend too long on this though!

Make a judgement on the effectiveness of the fieldwork data in developing your understanding by considering different factors and using available knowledge/evidence. **AO3 analysis and evaluation**

d  *Describe your fieldwork and **evaluate how effective** the data was in developing your understanding of the topic you investigated.

`8 marks`

### First paragraph – 'describe your fieldwork'

Read the exemplar below. Notice the following details were included: **title**, **location**, **sampling techniques**, **methods**, and **purpose of each method**.

*I went to the city centre of Birmingham to investigate the following question: 'How successful has the urban regeneration been of Birmingham's city centre?' I used systematic sampling to select fifteen people to do questionnaires in the city centre, asking people's opinions on the success of the regeneration. I also used random sampling to pick five sites to conduct environmental quality surveys to assess the quality of the surrounding environment.*

Write a description of your own human fieldwork below, including all the details listed above.

...................................................................................................................................................
...................................................................................................................................................
...................................................................................................................................................
...................................................................................................................................................

### Second paragraph – 'evaluate how effective your fieldwork data was in developing your understanding of the topic'

You need to discuss how **accurate** and **reliable** your data was, and consequently how confident you were in your results and conclusions, and how these shaped your understanding of relevant geographical theories or case studies. Use the boxes below to plan for this paragraph.

| How did your data effectively develop your understanding of the topic? | How did your data fail to effectively develop your understanding of the topic? |
|---|---|
| 1 | 1 |
| 2 | 2 |
| 3 | 3 |

### Concluding sentence

Link back to the question by answering 'how effective'. Use the sentence below to write your own concluding sentence.

*Overall, the fieldwork was quite/very effective in developing my understanding in* ............................
........................................... *However, it was not effective in developing my understanding of*

...................................................................................................................

...................................................................................................................

**Top tip:** The best answers will also discuss areas where your enquiry process was not as effective.

Now try writing an answer in full under timed conditions. Give yourself 11 minutes.

# Your own fieldwork (physical geography)

Complete the five boxes to help you evaluate your own physical fieldwork as you did for your human fieldwork.

**Stage 1:** In what ways was your fieldwork question SMART/not SMART?

> **Top tip:** Think about both your sampling strategy and methods to collect the data itself.

**Stage 2:** In what ways were your data collection methods accurate/not accurate? In what ways were your data collection methods reliable/not reliable?

> **Top tip:** Could you have improved your data presentation method or presented the data differently?

**Stage 3:** In what ways were your data presentation methods clear/not clear

> **Top tip:** Did your results support your hypotheses?

**Stage 4:** In what ways were your data effective/not effective in helping reach a conclusion?

> **Top tip:** Think about your methodology, data presentation and results.

**Stage 5:** What were the reasons you were confident/not confident in your conclusions? How was your understanding of the topic developed as a result of the investigation?

> **Top tip:** Link back to your original question and any supporting geographical theory.

---

**1** You will have taken part in fieldwork in a physical geography environment as part of your studies. Examples might include a river or a coastal area.

State your fieldwork question for investigation: ...................................................................

................................................................................................................................

Location of study area: ...............................................................................................

a   Describe how **one piece** of secondary information helped your study.    **1 mark**

...............................................................................

...............................................................................

> **Top tip:** Secondary information = information collected by someone other than you, e.g. climate data.

**b** Explain the sampling technique used for **one piece** of primary data collection. **2 marks**

........................................................................................................................

........................................................................................................................

........................................................................................................................

**Top tip:** For example, random, stratified, systematic or pragmatic.

Make a judgement by considering the arguments for and against. Justify your decision. **AO3 analysis and evaluation**

One paragraph on your data supporting the theory, one paragraph on it not supporting the theory.

**c** **To what extent** did your data support the theory behind your fieldwork question? **6 marks**

The key theory behind my fieldwork question was...

........................................................................................................................

........................................................................................................................

**Top tip:** Max one sentence.

Make your argument:

........................................................................................................................

........................................................................................................................

........................................................................................................................

**Top tip:** Learn some of your results, so you can use them here.

Counter your argument:

........................................................................................................................

........................................................................................................................

........................................................................................................................

**Top tip:** Any anomalies/trends that do not support the key theory?

**Conclude** in a sentence by using the 'opinion line' below to help you.

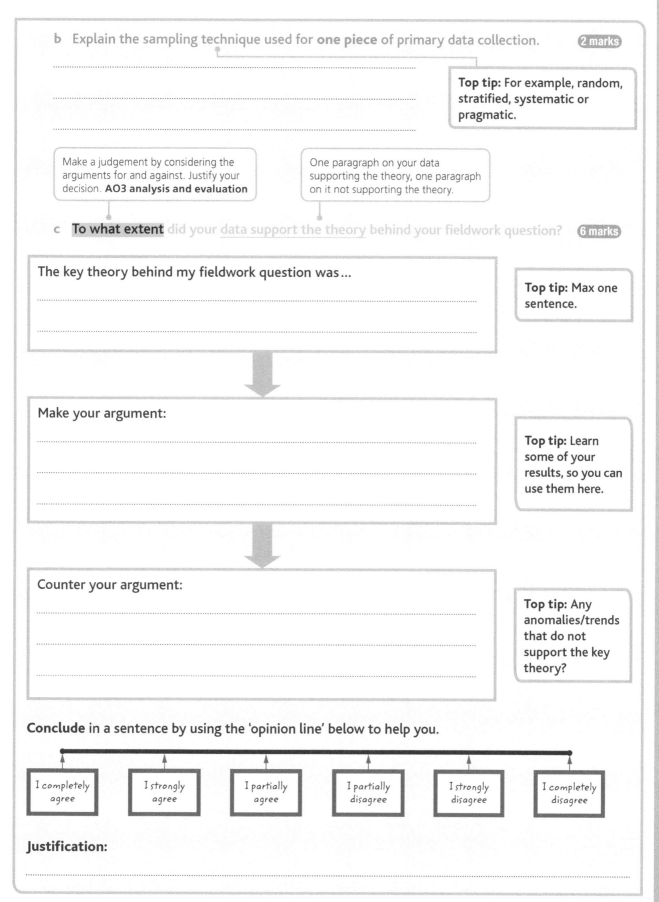

| I completely agree | I strongly agree | I partially agree | I partially disagree | I strongly disagree | I completely disagree |

**Justification:**

........................................................................................................................

## Key terms

'**Success**' is a common word in fieldwork questions. A 'successful' enquiry has methods that provide accurate and reliable data, so that you are confident in your conclusions, and therefore are able to answer your original question and improve your geographic understanding.

d **Evaluate to what extent** one <u>method</u> you used to collect your primary fieldwork data <u>was a success.</u>

(6 marks)

**Use the structure below to help you answer Question 1d.**

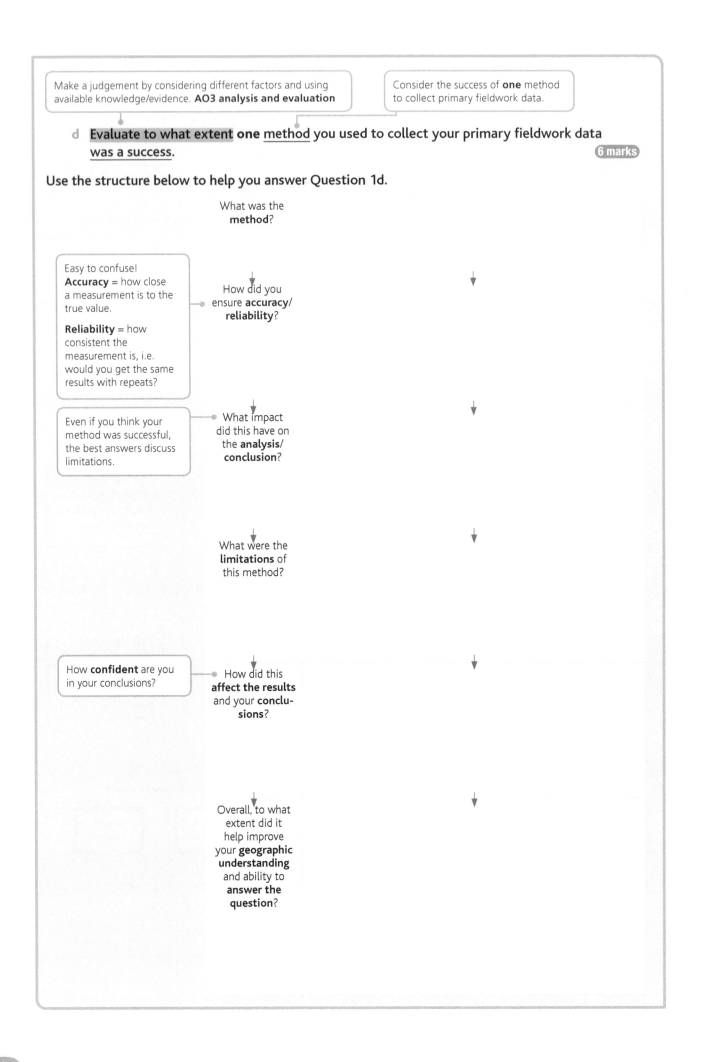

What was the **method**?

Easy to confuse!
**Accuracy** = how close a measurement is to the true value.

**Reliability** = how consistent the measurement is, i.e. would you get the same results with repeats?

How did you ensure **accuracy/ reliability**?

Even if you think your method was successful, the best answers discuss limitations.

What impact did this have on the **analysis/ conclusion**?

What were the **limitations** of this method?

How **confident** are you in your conclusions?

How did this **affect the results** and your **conclu- sions**?

Overall, to what extent did it help improve your **geographic understanding** and ability to **answer the question**?

> All 8 marks are for AO3 (**analysis and evaluation**). This question involves all six stages (see page 89) in the enquiry process. You need to reflect critically on any of Stages 1–4 to decide how successful they were in helping you reach a conclusion.

e * **Evaluate** how <u>successful</u> the <u>fieldwork process</u> was in helping you <u>reach a conclusion</u>. **8 marks**

**Use the radar chart to plan your answer. Mark an 'X' on each line to make a judgement on the three stages of the enquiry process.**

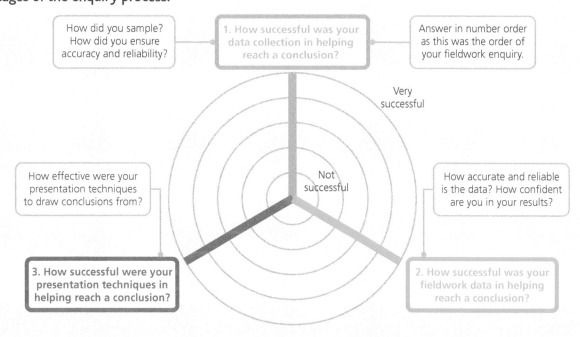

How did you sample? How did you ensure accuracy and reliability?

1. How successful was your data collection in helping reach a conclusion?

Answer in number order as this was the order of your fieldwork enquiry.

Very successful

Not successful

How effective were your presentation techniques to draw conclusions from?

3. How successful were your presentation techniques in helping reach a conclusion?

How accurate and reliable is the data? How confident are you in your results?

2. How successful was your fieldwork data in helping reach a conclusion?

**Now write an answer to this question in 11 minutes.**

..........................................................................................................................................................................
..........................................................................................................................................................................
..........................................................................................................................................................................
..........................................................................................................................................................................
..........................................................................................................................................................................
..........................................................................................................................................................................
..........................................................................................................................................................................
..........................................................................................................................................................................
..........................................................................................................................................................................
..........................................................................................................................................................................
..........................................................................................................................................................................
..........................................................................................................................................................................
..........................................................................................................................................................................
..........................................................................................................................................................................

**1** **This workbook will help you** prepare for your OCR GCSE (9–1) Geography A exam.

**2** **Build your skills** and prepare for every question in the exam using:
- clear explanations of what each question requires
- short answer activities that build up to exam-style questions
- spaces for you to write or plan your answers.

**3** **Answering the questions** will help you build your skills and meet the assessment objectives (AOs):

**AO1**: remembering geographical facts

**AO2**: understanding geographical concepts and processes

**AO3**: evaluating evidence or using evidence to make a decision

**AO4**: using skills to investigate maps and graphs or making calculations.

**4** **You still need to** read your textbook and refer to your revision guide and lesson notes.

**5** **Answers** to every question in the book are available at **www.hoddereducation.co.uk/workbookanswers**

The Publishers would like to thank the following for permission to reproduce copyright material.

**Photo credits**

**p.14** © Matthew Fox; **p.15** © RichSTOCK / Alamy Stock Photo; **p.23** © Mazur Travel - stock.adobe.com; **p.26** © Dennis Hardley / Alamy Stock Photo; **p.28** © travelbild / Alamy Stock Photo; **p.30** © Kateryna - stock.adobe.com; **p.39** © Robert Evans / Alamy Stock Photo; **p.45** © alpegor - stock.adobe.com; **p.66** © Heritage Image Partnership Ltd / Alamy Stock Photo; **p.80** *t* © AGAMI Photo Agency / Alamy Stock Photo; *b* © robertharding / Alamy Stock Photo

**Acknowledgements**

Figure 2.6 on p.35 contains public sector information licensed under the Open Government Licence v3.0.

OS maps on p.11 and p. 75 reproduced from Ordnance Survey mapping with permission of the Controller of HMSO. © Crown copyright and/or database right. All rights reserved. Licence number 10003470. Ordnance Survey (OS) is the national mapping agency for Great Britain, and a world-leading geospatial data and technology organisation. As a reliable partner to government, business and citizens across Britain and the world, OS helps its customers in virtually all sectors improve quality of life.

Map on p.76 powered by TfL Open Data.

Every effort has been made to trace all copyright holders, but if any have been inadvertently overlooked, the Publishers will be pleased to make the necessary arrangements at the first opportunity.

Although every effort has been made to ensure that website addresses are correct at time of going to press, Hodder Education cannot be held responsible for the content of any website mentioned in this book. It is sometimes possible to find a relocated web page by typing in the address of the home page for a website in the URL window of your browser.

Hachette UK's policy is to use papers that are natural, renewable and recyclable products and made from wood grown in well-managed forests and other controlled sources. The logging and manufacturing processes are expected to conform to the environmental regulations of the country of origin.

Orders: please contact Hachette UK Distribution, Hely Hutchinson Centre, Milton Road, Didcot, Oxfordshire, OX11 7HH.

Telephone: +44 (0)1235 827827. Email education@hachette.co.uk Lines are open from 9 a.m. to 5 p.m., Monday to Friday. You can also order through our website: www.hoddereducation.co.uk

ISBN: 978 1 5104 6050 8

© Matthew Fox 2019

First published in 2019 by
Hodder Education,
An Hachette UK Company
Carmelite House
50 Victoria Embankment
London EC4Y 0DZ

www.hoddereducation.co.uk

Impression number  10 9 8 7 6 5 4

Year        2023 2022 2021

Cover photo © Jim Zuckerman/Corbis

Illustrations by Aptara Inc.

Typeset in India by Aptara Inc.

Printed in the UK

A catalogue record for this title is available from the British Library.

Ordnance Survey
The world's trusted geospatial partner

**HODDER EDUCATION**
t: 01235 827827
e: education@hachette.co.uk
w: hoddereducation.co.uk

ISBN 978-1-5104-6050-8
9 781510 460508

MIX
Paper | Supporting responsible forestry
FSC™ C104740